Fluid Power
Educational
Series

I0491737

Hydraulic Cylinders

(In the SI Units)

Joji Parambath

Hydraulic Cylinders
(In the SI Units)

Copyright © 2022 Joji Parambath

ISBN: 9798653721243

https://fluidsys.org

First Edition: 2020
Revised Edition: 2022

Disclaimer of Liability

The contents of this book have been checked for accuracy. Since deviations cannot be precluded entirely, we cannot guarantee full agreement. Only qualified personnel should be allowed to install and work on pneumatic and hydraulic equipment. Qualified persons are defined as persons who are authorised to commission, ground, and tag circuits, equipment, and systems following established safety practices and standards.

Dedicated to

all my teachers

Table of Contents

Chapter	Description	Page No
--	Preface	vii
1	Introduction and Basic Cylinder Working	1
2	Terms and Definitions - Hydraulic Cylinders	3
3	Principal Parts and Body Styles of Hydraulic Cylinders	11
4	Side loads in Hydraulic Cylinders	22
5	Piston-rod Buckling	23
6	Classification and Types of Hydraulic Cylinders	25
7	Position Transducers for Hydraulic Cylinders	36
8	Installation and mounting of Hydraulic Cylinders	39
9	Advantages of Hydraulic Cylinders	42
10	Applications Notes, Hydraulic Cylinders	43
11	Standards, Hydraulic Cylinders	45
12	Maintenance and Safety of Hydraulic Cylinders	46
13	Design of Hydraulic Cylinders	51
14	Manufacturing Process of Hydraulic Cylinders	53
15	Objective Type Questions	66
16	Review Questions	67
17	Numerical Problems	68
Appendix 1	Standard bore diameters and piston–rod diameters in the SI units	70
Appendix 2	Theoretical Cylinder Forces	71
Appendix 3	Important Specifications to be considered while Selecting Hydraulic Cylinders	72
Appendix 4	Different types of Piston Seals, Ports, Mounting Styles	73
Appendix 5	Seal materials and their temperature Ranges	74
18	References	75

PREFACE

Hydraulic cylinders are simple, low-cost, and easy-to-install devices that are ideal for generating powerful linear movements. Manufacturers are bringing out various types of actuators with innovative features to make them more reliable, efficient, and safe. The latest industrial hydraulic cylinders can incorporate sensor feedback and electro-hydraulic servo valves for sophisticated speed control and position control of the associated loads.

The book is an attempt to bring out the essential technical information related to hydraulic cylinders, in a simple and easy to understand manner. The topics are logically arranged for a simple to the sophisticated level progression of the subject matter. Even though there is no end to the advancement of the technology, a strong foundation in hydraulic technology will sure to help the reader for a quick understanding of the future developments in the field. The book uses the SI system of units.

Many other fluid power topics are given in other textbooks under the fluid power educational series by the same author. A list of all the books is given at the end of the book (Page No. 78). Also, please see the details at: https://jojibooks.com.

Enjoy reading the book.
Your feedback is most welcome.

JOJI Parambath

Chapter 1 | Introduction and Basic Cylinder Working

Introduction
A hydraulic actuator is a positive displacement device used in a hydraulic system to drive the attached load to get some useful work. Its primary function is to convert hydraulic power into mechanical power. The resulting output motion can be either linear or rotary. Accordingly, there are two basic types of hydraulic actuators. They are: (1) Linear actuators and (2) Rotary actuators. The linear actuators convert hydraulic energy into straight-line mechanical energy. Next, the rotary actuators convert hydraulic energy into rotary mechanical energy. An example of the linear actuator is a cylinder, and the rotary actuator is a hydraulic motor.

Linear Actuators
A linear hydraulic actuator, as shown in Figure 1.1, converts hydraulic power into a controllable linear force or motion or both. Technically and economically, hydraulic and pneumatic cylinders are the optimum form of linear actuators. However, a cylinder used in the high-pressure hydraulic system can provide a much higher force output as compared to that provided by a cylinder of the comparable size used in a pneumatic system.

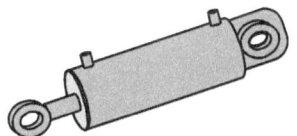

Figure 1.1 | Hydraulic cylinder

In general, the cylinders account for the majority of actuators used in fluid power systems. Hydraulic actuators are available in a gamut of sizes, types, body styles, and mounting configurations to meet varied requirements of applications.

Working Principles of a Hydraulic Cylinder

Figure 1.2 shows the cross-sectional view of a typical hydraulic cylinder. It primarily consists of a barrel, piston-and-piston-rod assembly, end-caps, and necessary seals and ports. The end-caps are securely fastened to the barrel. The piston with tight sealing forms two fluid chambers (piston chamber and piston-rod chamber) and can move within the barrel having two bearing surfaces. The cylinder is provided with two ports for the entry or exit of the system fluid. They are: (1) piston-side (cap-end) port X and (2) rod-side (head-end) port Y.

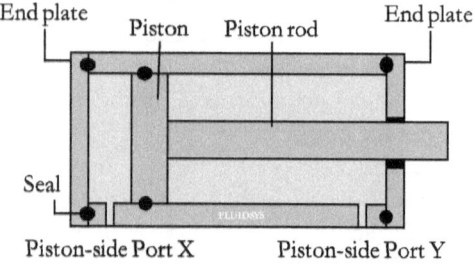

Figure 1.2 | A schematic diagram of a basic hydraulic cylinder

If the system fluid is pumped into the piston chamber through the port X and the fluid in the piston-rod chamber is discharged through the port Y, then the piston-and-rod assembly extends with a definite force (thrust). If the system fluid is pumped into the piston-rod chamber through the port Y and the fluid in the piston chamber is discharged through the port X, then the piston-and-rod assembly retracts with a definite force (pull). This type of hydraulic cylinder is called a 'double-acting' cylinder, as the extension and retraction of the cylinder are obtained hydraulically.

If the motion of the hydraulic cylinder is obtained hydraulically in only one direction, then such a cylinder is called a 'single-acting' cylinder. The motion in the opposite direction can then be obtained by a spring or gravity or by any other external force.

Chapter 2 | Terms and Definitions -Hydraulic Cylinders

Some essential parameters concerned with the operation and applications of hydraulic cylinders are their bore diameter, piston-rod diameter, force (thrust and pull), stroke length, speed, and piston-rod buckling. The following sections describe these terms:

Maximum Operating Pressure (P): It is the pressure that overcomes all resistances in the system, which includes both useful work and losses. Alternatively, it is the maximum working pressure that the cylinder can sustain without adverse consequences.

Bore Diameter (D): It refers to the diameter at the bore of the cylinder (See Figure 2.1). It can be used to calculate the bore area of the cylinder. It is also equal to the piston diameter, in a close-fitting hydraulic cylinder.

Piston-rod Diameter (d): It refers to the diameter of the piston-rod of the cylinder (See Figure 2.1).

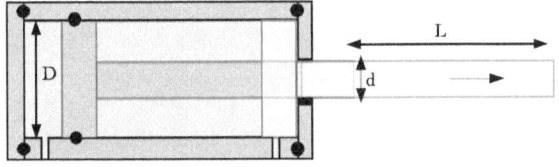

Figure 2.1 | Cylinder parameters

Stroke Length (L): It is the distance through which the piston-and-piston-rod assembly of the cylinder moves through the cylinder.

Note: Details of standard bore diameters and piston-rod diameters are given in Appendix 1.

Maximum Stroke Length: It is the maximum linear movement that a cylinder can produce. For standard models of double-acting cylinders, the maximum stroke lengths can be up to 2000 mm, and for special designs, the stroke lengths can be up to 6000 mm.

Cylinder Thrust/Pull (F): The theoretical thrust (F) during the forward stroke or pull (F) during the return stroke of the cylinder can be determined by multiplying the effective area (A) of the piston by the working pressure (P) to which it is subjected, according to Pascal's law. The active area (A_{ext}), considered for the calculation of the cylinder thrust, is the full area (A_p) of the cylinder bore and is given by ($\pi.D^2/4$). The parameter 'D' denotes the piston (bore) diameter. Further, the active area (A_{ret}), considered for the calculation of the cylinder pull, is the area (A_p) of the piston minus the piston-rod area (A_r), and it is given by the expression [$\pi. (D^2 - d^2)/4$]. The parameter 'd' denotes the piston-rod diameter. The theoretical thrust and the theoretical pull are given by:

Thrust, F (Newton)	= P (Pascal) x A_{ext} (m²)
Pull, F (Newton)	= P (Pascal) x A_{ret} (m²)

Where,

A_p is the piston area
A_r is the piston-rod area
A_{ext} is the active area during extension: ($A_{ext} = A_p$)
A_{ret} is the active area during retraction: ($A_{ret} = A_p - A_r$)

Table A2.1 in Appendix 2 gives the theoretical forces of hydraulic cylinders in the SI system of units. These figures do not account for the seal or packing friction in these cylinders. This type of friction is estimated to affect the thrust of the cylinders by about 10%.

Limitations on Maximum Thrust Force

The maximum thrust force that a cylinder can practically provide is limited by its piston-rod diameter and overall length. In cylinders with longer piston-rods, the piston-rod must be able to handle the thrust forces generated by the application. The cylinder must also be supported adequately.

Note that a head-end mounting provides greater column strength than the cap-end mounting, due to the smaller distance between the mounting points in the head-end mounting than that in the cap-end-mounting.

The piston-rod size of a hydraulic cylinder can be selected from the size charts, with the help of values of its free buckling length and the load imposed on the cylinder.

Example 2.1: A high-pressure double-acting hydraulic press cylinder with an effective piston area of 71 cm² for push stroke, and a piston-rod area of 22 cm², operating at 700 bar does produce what theoretical forces for the push stroke and pull stroke?

Solution

Piston area, push stroke, A_{push} = 71 cm²
Piston Rod area, A_{rod} = 22 cm²
Pressure, P = 700 bar

Effective piston area, pull stroke, A_{pull} = A_{push} - A_{rod}
= (71 -22) cm² = 49 cm²

Thrust, F_{push} = P x A_{push}
= (700x10⁵) x 71 x 10⁻⁴
= 497000 N

Pull, F_{pull} = P x A_{pull}
= (700x10⁵) x 49 x10⁻⁴
= 343000 N

Cylinder Input Power: The input power (P_{input}) supplied to the hydraulic cylinder, in the SI system units is given below:

$$P_{input} \text{ (Watts)} = P \text{ (Pa)} \times Q_A \text{ (m}^3/\text{s)}$$

Cylinder Output Power: The mechanical output power (P_{output}) of the hydraulic cylinder in the SI system of units is given below:

$$P_{output} \text{ (Watts)} = \text{Force (N)} \times \text{Velocity (m/s)}$$

The input power supplied must be higher than the required output power to make allowances for losses on account of friction and leakage.

Cylinder Speed: Assume that the piston-rod assembly of a cylinder moves with a velocity of 'v' when pushed by the system fluid with a flow rate 'Q'. Further, assume that the cylinder piston of area 'A' has moved a distance 'S' in time 't' for attaining the velocity v.

Figure 2.2 (a) shows the working position of the cylinder with the piston in position 1. Figure 2.2 (b) shows the working position of a cylinder with the piston in position 2 with the positions '1' superimposed. Mathematically,

$$v = S/t \qquad \text{or} \quad t = S/v$$

We can easily relate the theoretical flow rate (Q) of the system fluid to the speed (v) at which the piston-rod moves if we consider the cylinder volume (V) that must be filled with the fluid and the distance (S) through which the cylinder piston must travel at the specified speed.

The volume (V) of the cylinder is the length of the stroke (S) multiplied by the piston area (A). The following section gives the flow rate (Q) to achieve the required speed (v), in the SI system units.

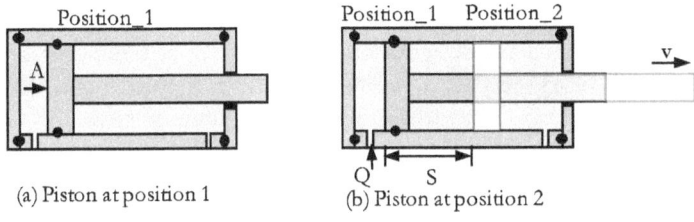

Figure 2.2 | Illustration of a cylinder in two piston positions

$$Q \ (\mathrm{m^3/s}) = \frac{V \ (\mathrm{m^3})}{t \ (\mathrm{s})} = \frac{A \ (\mathrm{m^2}) \ \mathrm{x} \ S \ (\mathrm{m})}{t \ (\mathrm{s})} = A \ (\mathrm{m^2}) \ \mathrm{x} \ v \ (\mathrm{m/s})$$

It can be observed from the equation mentioned above that the speed (v) of a given cylinder depends on the flow rate (Q) of the system fluid. That is, a small-diameter cylinder moves faster as compared to a large-diameter cylinder with the flow rate remaining the same.

Examination of this equation also shows that the double-acting cylinder tends to produce somewhat a higher speed when retracting than when extending, provided that the flow rate remains the same. This speed difference is mainly due to the different active areas exposed to the system fluid.

Cylinder Drift
Under ideal conditions, a hydraulic cylinder should hold its position when stopped. However, it tends to drift from its intended position, at times. The cause of the drifting can often be attributed to the associated control valve rather than the cylinder itself. If the valve spool is blocked entirely in its neutral position, the cylinder should maintain its position. However, if the valve spool permits leaks in its neutral position, possibly due to a damaged seal, both working ports of the valve may be connected to the system reservoir, and consequently, the cylinder may drift from its intended position.

Operating Temperature
Seal compounds are designed for normal operating temperature from -20°C to +80°C. The operating temperature should not exceed 45°C to 50°C to ensure the long life of the system.

Types of Hydraulic Loads: There are three types of loads associated with hydraulic systems. They are: (1) Resistive load (Positive load), (2) Overrunning load (Negative load), and (3) Inertial load.

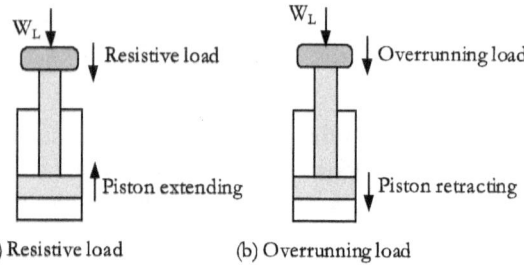

(a) Resistive load   (b) Overrunning load

Figure 2.3 | Types of loads associated with hydraulic systems

A resistive load on a hydraulic actuator tends to oppose the motion of the actuator, as shown in Figure 2.3(a). The cutting or shearing operation acting against the motion of the cylinder is an example of the resistive load.

An overrunning load moves and acts in the same direction as that of the associated actuator. A descending heavy weight on a cylinder, as shown in Figure 2.3(b), is an example of an overrunning load.

An inertial load is a load that tends to resist the acceleration or deceleration of the actuator to which it is connected. A heavy flywheel or fan attached to the shaft of a motor is an example of an inertial load.

Summary of Relations for Hydraulic Cylinders

Figure 2.4 gives the summary of essential relations of hydraulic cylinders, in the SI system units for easy understanding and the correlation of these relations by the reader.

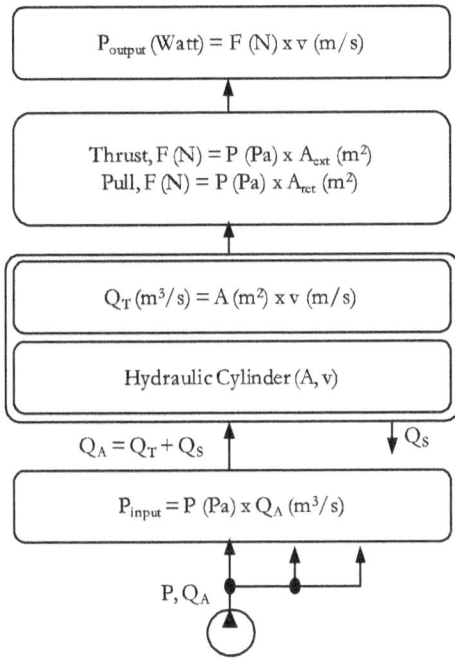

Figure 2.4 | Summary of relations - hydraulic cylinders

Example 2.2: Determine the thrust of a hydraulic cylinder with a bore diameter of 100 mm, used for a flyover levelling application, and operating at a pressure of 700 bar?

Solution

Bore diameter, D	= 100 mm
Operating pressure, P	= 700 bar

Piston area, extension stroke, $A = \pi D^2/4$
$$= \pi \times (100 \times 10^{-3})^2 /4$$
$$= 0.00785 \text{ m}^2$$

Thrust, F
$$= P \times A$$
$$= 700 \times 10^5 \times 0.00785$$
$$= 549500 \text{ N}$$

Example 2.3: A double-acting hydraulic clamping cylinder must move out with a velocity of 0.5 m/s during the extension stroke. Calculate the flow rate demanded by the cylinder that is operating at a pressure of 207 bar and producing a thrust of 50 kN.

Solution

Thrust, F	= 50000 N
Pressure, P	= 207 bar
Speed, v	= 0.5 m/s

Area, A_{ext}
$$= F/P$$
$$= 50000/ (207 \times 10^5)$$
$$= 0.002415 \text{ m}^2$$

Flow rate, Q
$$= A_{ext} \times v$$
$$= 0.002415 \times 0.5$$
$$= 0.0012 \text{ m}^3/s$$

Chapter 3 | Principal Parts and Body Styles of Hydraulic Cylinders

Hydraulic cylinders are designed to operate at high pressures and handle high loads in demanding operating conditions. Therefore, they need to be constructed with materials of high strength, expert workmanship, and advanced features to provide ruggedness, high quality, maintenance-free operation, and long service life.

A hydraulic cylinder is a combination of essential and optional parts, such as a barrel, a piston, a piston-rod, end caps, cushion seals, wear bands, a piston-rod seal/wiper, piston-rod bearings, piston-rod boots, and a stop tube. A hydraulic cylinder may incorporate many advanced features, such as a magnetic piston, and end-of-stroke or mid-stroke sensors, to sense the position of the piston economically and reliably. Air bleeds can also be provided to exhaust the air accumulated in the cylinder. Figure 3.1 shows the cross-sectional view of a simple double-acting hydraulic cylinder.

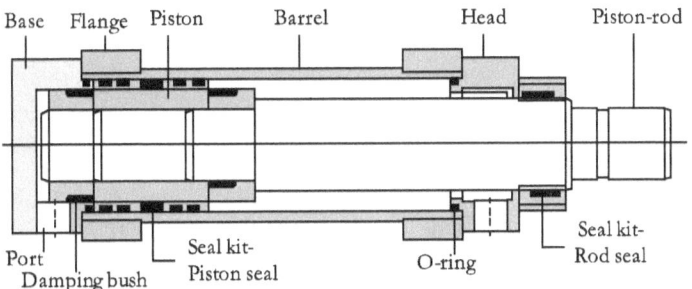

Figure 3.1 | A cross-sectional view of a hydraulic cylinder

It may be noted that a cylinder with leak-proof piston seals and piston-rod seals provide consistent performance at very slow speeds against a wide range of load resistances.

Barrel

The barrel of a hydraulic cylinder is made of a high-strength seamless-drawn tube that is precision-machined to a perfect finish. The internal surface of the barrel must be very smooth so that the wear and leakage in the cylinder can be controlled. A good quality manufacturing process of the barrels ensures smooth surface finish, and the straightness and roundness of cylinders, which in turn, guarantees their smooth action and superior fluid sealing property. An aluminium tube or a brass tube, or a steel tube with the hard-chromed bearing surface (Figure 3.2) can be used as part of the barrel of a hydraulic cylinder if it is likely to be exposed to corrosive fluids. These materials are also better conductors of heat, and, therefore, they assist in removing the heat from applications with high-frequency cyclic operations. However, when using the aluminium body for the barrel, the cylinder should not be subjected to any shock or high-inertia loading.

Figure 3.2 | Tubes for cylinder barrel

Piston

Figure 3.3 shows the schematic diagram of a piston. The primary function of the piston in a hydraulic cylinder is to transmit the force to the load attached to its piston-rod. Apart from this, it acts as the bearing in the cylinder barrel.

The piston must be a perfect fit inside the cylinder barrel. It must be reasonably cylindrical and finely finished for its smooth output motion. It is made of fine-grained alloy steel with a coating of bronze. It can be threaded or welded to the piston-rod. The grooves on the piston are provided to contain the packing or seals. For a heavy-duty application, an additional bearing ring may be fitted to the piston for its excellent wear resistance.

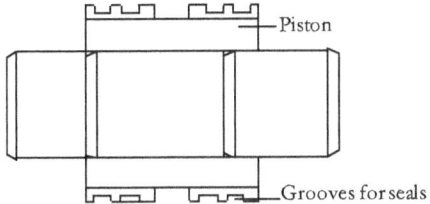

Figure 3.3 | Schematic diagram of a cylinder piston

Piston-rod
The piston-rod of a hydraulic cylinder moves in and out of the cylinder barrel and comes into contact with its surrounding atmosphere. A surface, that is smooth, hard, and corrosion-resistant, is essential for the outer surface of the piston-rod. Therefore, the piston-rod is, usually, made of induction-hardened steel or stainless steel (Figure 3.4). It may also be chrome-plated with an ultra-fine surface finish to ensure its resistance to wear and corrosion. Good slide rings are provided in the piston-rod for its excellent wear resistance and proper sealing. A piston-rod may be of a hollow design or externally guided, or both.

Figure 3.4 | Piston rod

End-caps
The end-caps are shown in Figure 3.5. They are attached to the ends of the cylinder barrel to enclose the pressure chamber in the cylinder. They are cast from iron or aluminium or made from high-quality steel. They may be designed with square or round

shapes to match with the barrel shape. They can be fixed by tie-rods, or threaded or welded to the barrel. They also incorporate threaded entries for ports. They have to withstand any bending stress and shock loads to which they may be exposed. The end-of-travel shocks in a cylinder can be absorbed with the cushion valves built into its end-caps.

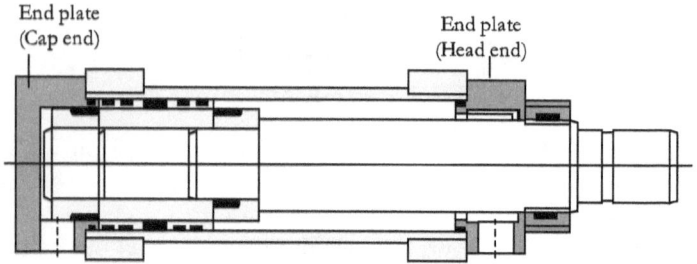

Figure 3.5 | End caps attached to the barrel

Cushion

It is an optional device in a hydraulic cylinder that is incorporated at one end or both ends to minimise the shock loads as the piston approaches its end-of-stroke position. A hydraulic cylinder should be provided with cushions when the piston velocity is expected to be over 0.1 m/s. The details of cushion cylinders are given on Page 29.

Seals

A small amount of leakage up to 0.05 lpm across the piston of a hydraulic cylinder is considered normal. A large amount of leakage can be a severe problem leading to energy wastage, loss of efficiency, environmental fouling, and safety hazards. Leakage in a cylinder can be controlled by using appropriate sealing devices (Figure 3.6). A seal is an elastomeric part used in a hydraulic component to close the imperfections in their mating surfaces. The installation of proper seals in the cylinder helps maintain the pressure of the associated system, prevent the loss of fluid from the system, reduce friction, and keep the

14

contamination out of the system. However, it may be remembered that no sealing device can be 100% positive.

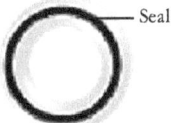

Figure 3.6 | Hydraulic seals

Sometimes a seal for a hydraulic cylinder may be deliberately designed to allow a controlled amount of internal leakage for providing the lubrication of its internal moving parts. A good seal must be compatible with many types of fluids and must be capable of withstanding the harsh industrial environment.

Piston Seals
Piston seals, as shown in Figure 3.7, are fitted into the grooves of the cylinder piston. Filled PTFE lip seals with internal stainless steel springs can be used as piston seals to optimise seal performance.

Pressure in the cylinder makes the lip of a cup seal spread out and grip the barrel tightly to give a positive sealing. However, it is capable of sealing only in one direction. Therefore, double cup seals must be used in double-acting cylinders to provide sealing in both directions.

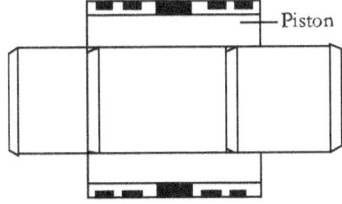

Figure 3.7 | Piston seals

Wear Bands

Wear rings/bands, as shown in Figure 3.8, in a hydraulic cylinder serve to limit the wear on its piston and piston-rod seals, and provide superior protection against side loads. They are, usually, made from glass-reinforced nylon, which has excellent wear resistance and non-scoring properties.

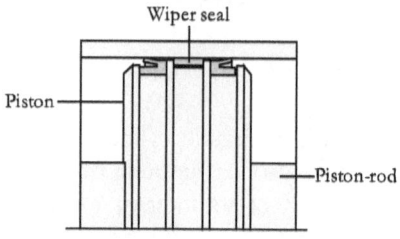

Figure 3.8 | Wear bands

Piston-rod Seal/Wiper

Figure 3.9 shows the piston-rod seal and wiper seal. These seals can be energised with fluorocarbon O-rings to maintain consistent contact pressure with the piston-rod. The specially-designed piston-rod seal prevents the pressurised fluid in a hydraulic cylinder from leaking out even as it allows the piston-rod to move freely back and forth through the piston-rod gland. The wiper/scraper part of the seal prevents external contamination, such as dust and dirt, from entering the cylinder through the piston-rod gland.

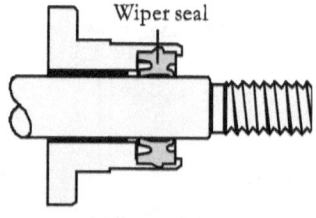

Figure 3.9 | Piston-rod seal/Wiper

The seal also prevents damage to the bearing, seals, and piston-rod. It consists of a honed wiping element of age-hardened beryllium copper surrounded by a synthetic rubber elastomer part. During the operation, the beryllium copper ring polishes the rod without any damage to its surface finish.

Piston-rod Bearing
The piston-rod bearing in a hydraulic cylinder guides the piston-rod as it passes back and forth through the rod gland. It also supports the weight of the piston-rod and the load attached to the piston-rod. Piston-rod-end bearings are made of brass or bronze to take care of the side loads on the piston-rod and ensure proper lubrication.

Piston Magnet
The piston of a cylinder can be provided with a magnet for sensing the piston position economically and reliably. By incorporating sensing capability into cylinders, the need to install stand-alone proximity sensors can be eliminated.

Air Bleeds
Hydraulic cylinders are considered self-bleeding when cycled full stroke. However, air bleeds can be recessed into one end or both ends of the cylinder body for venting trapped air in the cylinder. For bleeding air, hydraulic pressure is typically set at 10 to 30 bar. First, move the piston to the end-of-stroke position, and then slightly open the bleeder screw until bubble-free oil emerges. Then tighten the bleeder screw again. Bleed the other side of the cylinder in the same way.

Piston-rod Boots
When a cylinder is used in a highly contaminated environment, its exposed piston-rod should be covered with a boot or bellows (Figure 3.10). The piston-rod boot is a collapsible cover made from a neoprene-coated fabric, which is impervious to oil, grease, and water. This arrangement is made to protect its piston-rod, seals, and bearing against contamination.

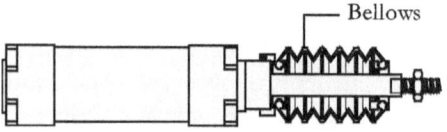

Figure 3.10 | Piston-rod boots

Stop tube

The stop tube is used in a horizontally-mounted hydraulic cylinder, especially with a long-stroke, poorly-guided piston-rod. Figure 3.11 shows the schematic diagram of the stop tube. It is a separator between the head-end cap and the piston of the cylinder, intended to prevent the piston from reaching the end of the cylinder barrel. This separation reduces the moment forces developed between the piston-rod bearing and the piston when the rod is fully extended. The reduction of the moment forces helps to maintain the bearing loads within the acceptable limits and gives the cylinder a higher side-load capacity and stability when compared to that of a regular cylinder without a stop tube.

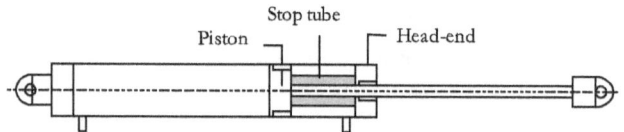

Figure 3.11 | A hydraulic cylinder with a stop tube

Body Styles of Hydraulic Cylinders

Hydraulic cylinders are manufactured in a variety of body styles according to the way their outer parts are assembled. Based on their assembly types, there are four basic types of hydraulic cylinders. They are: (1) Tie-rod cylinders, (2) Mill type cylinders, (3) Threaded-end cylinders, and (4) Welded cylinders. Any of these body styles can be selected according to the nature of the application and the surrounding environment.

Tie-rod Hydraulic Cylinders

This type of cylinder is used most often in industrial hydraulic applications. In a tie-rod cylinder, four or more high-strength threaded steel tie-rods run through the length of the cylinder. The tie-rods actively hold the barrel and two end-caps together. The size and strength of the tie-rods and their fasteners determine the strength of the cylinder. High-tensile tie-rods in a cylinder commonly bear a large portion of the applied load and absorb internal energy stresses. A large-bore high-pressure tie-rod cylinder may have as many as 24 tie-rods. Figure 3.12 shows the front and end views of a tie-rod cylinder.

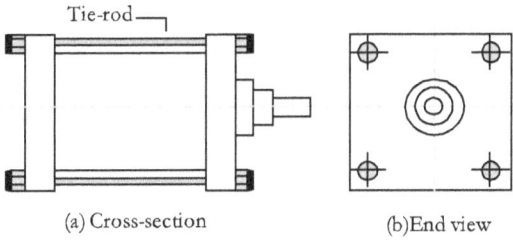

(a) Cross-section   (b)End view

Figure 3.12 | Front and end views of a tie-rod cylinder

The tie-rod cylinders have the advantages of rugged design, low maintenance, and long service life. They can be completely disassembled for service and repair. Therefore, heavy-duty cylinders used in the automotive, mobile and machine tool industries are of the tie-rod design. These sturdy and dependable cylinders can also be used in various light-duty and medium-duty hydraulic applications.

However, long-stroke tie-rod cylinders pose several problems, such as tie-rod stretch, tie-rod sag, and cylinder flexing and twisting. The long steel tie-rods can stretch to a certain degree under severe tension, and they may stretch to such an extent that the barrel separates from the end-caps. A long steel tie-rod also sags due to its weight. Therefore, it must be fitted with one or more intermediate supports along the cylinder barrel to prevent

19

it from sagging. The tie-rod cylinders are also susceptible to the problems of twisting and flexing. Manufacturers publish pressure limits and mounting style restrictions for cylinders because of the tie-rod stretch and flex and twist problems inherent in the tie-rod design.

Mill Type Hydraulic Cylinders
They are heavy-duty cylinders designed for harsh environments and demanding service conditions, as in steel mills, mines, and furnaces. A mill type cylinder, usually, has a thick-wall tubular housing, which provides exceptional strength and heat resistance. A cylinder of this type has its end-caps fastened to the mating flanges with bolts or cap screws, as shown in Figure 3.13.

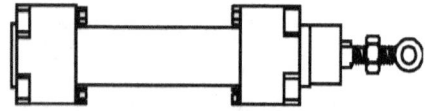

Figure 3.13 | Mill type hydraulic cylinder

Mill type cylinders are built to withstand buckling under compressive loads and can be used in corrosive and non-corrosive environments. Apart from the applications in steel mills, furnaces, and mines, they are also well-suited for other applications as in offshore and metal-forming industries.

Welded Hydraulic Cylinders
Figure 3.14 shows a welded hydraulic cylinder. The heavy-duty barrel of this type of cylinder is welded along the entire outside diameter to each of its end-caps. Also, its ports are welded to the barrel, and the mountings are welded to the cylinder body. The welded cylinders provide an incredibly strong bond and high strength. They are also very compact.

Moreover, they have smooth exterior surfaces, free of tie-rods and fasteners. These constructional features mean that there are fewer places for dirt and debris to accumulate. The round body is free of obstructions, allowing it to be cleaned easily.

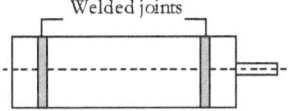

Figure 3.14 | A welded hydraulic cylinder

The welded cylinders are compact and usually designed for high strength. They are space-efficient in their overall length, as compared to the tie-rod cylinders. They are also not susceptible to the problems of tie-rod sag and stretch. Because of these advantages, they dominate the mobile hydraulic equipment market and the heavy equipment industry. However, welded cylinders have the disadvantage that they cannot be disassembled for servicing.

Threaded-end Hydraulic Cylinders
In this type of cylinder construction, the end-caps of the cylinder are threaded to fit the internal thread of the tubular cylinder bore. In applications where space is an important consideration, such cylinders can be used. The food-processing and food-packaging industries make extensive use of this type of cylinder body style.

Note: Appendix 4 gives different types of piston seals, ports, mounting styles. Appendix 5 gives seal materials and their temperature ranges.

Chapter 4 | Side Loads in Hydraulic Cylinders

The side load in a cylinder is the force component that acts laterally across the axis of its bearings, piston-rod seals, and piston. A long-stroke hydraulic cylinder often experiences significant side load forces while in operation, especially when it is fully extended.

Many reasons can be attributed to the development of side loads in a cylinder. An off-centre load on the cylinder, as shown in Figure 4.1, causes side loads on its bearings, seals, and piston. Similarly, the cylinder experiences side loads if it is not aligned or mounted correctly. Even the weight of the cylinder or its piston-rod exerts side load on the rod bearings and piston bearings.

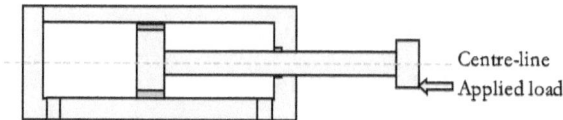

Figure 4.1 | A hydraulic cylinder applied with an off-centre load.

The side loads on a hydraulic cylinder are liable to cause stress on its bearings, seals, piston, mounting flange, and mounting bolts. They ultimately reduce the service life of the cylinder.

It is essential to reduce the side loads imposed on long-stroke cylinders. It can be reduced by the proper mounting arrangement of the cylinder. The axis of the cylinder must be correctly aligned to the axis of the load attached.

The side load capacity of a cylinder can be improved by providing adequate bearing support, heavy-duty wear ring, oversized piston-rod, dual pistons, and/or stop tube.

Chapter 5 | Piston-rod Buckling

Hydraulic cylinders capable of giving very long strokes are essential for some applications. However, if a compressive axial load is to be applied to the piston-rod of such a long-stroke cylinder, it must be within the safety limit to prevent buckling of the piston-rod.

Piston-rod buckling can occur when the piston-rod bends under a load. In other words, the buckling can occur in the cylinder if the piston-rod is not sized to match its stroke and the connected load. Further, the buckling of the piston-rod is likely to occur if it is longer and thinner. Figure 5.1 is a schematic diagram illustrating the piston-rod column failure in a hydraulic cylinder.

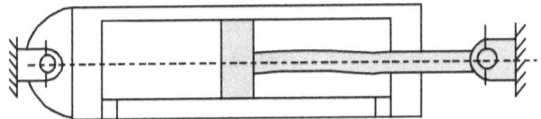

Figure 5.1 | Illustration of piston-rod column failure in a hydraulic cylinder

Due to the buckling stress, the permissible load that can be applied on the piston-rod of a long-stroke hydraulic cylinder should be less than that ought to have been permitted by a short-stroke cylinder of the same piston area used under identical operating conditions. The permissible load on the cylinder should not go beyond a specific value as given by the Euler's and Tatmajer formulae.

According to Euler's formula:

$$F_k = \frac{\Pi^2 E J}{S \, l_k^2} \qquad \text{If } \lambda > \lambda_g$$

According to Tatmajer formula:

$$F_k = \frac{\Pi (335 - 0.62 \lambda) \, d^2}{4 S} \qquad \text{If } \lambda <= \lambda g$$

Where,

F_k = Permissible buckling force (N)
E = Modulus of elasticity
= 2.1×10^5 N/mm^2 for carbon steel
J = Moment of inertia (mm^4) = $\pi d^4 / 64 = 0.0491 \, d^4$
d = Piston-rod diameter in mm
S = Safety factor (chosen as 3.5 to 5)
l_k = Equivalent free buckling length in mm
= (1 to 2) x stroke length, l
(Depending on the mounting styles shown in Figure 5.2)
λ = Slenderness ratio = $4 \times l_k / d$
$\lambda_g = \pi \times \sqrt{(E / 0.8 \times R_e)}$
R_e = Yield strength of the piston-rod material

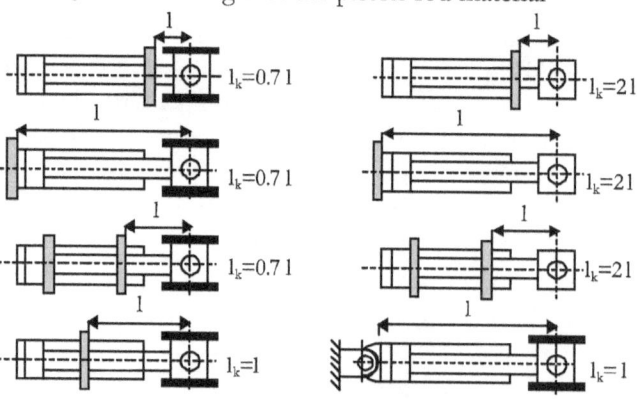

Figure 5.2 | Typical free buckling lengths for different mounting styles of hydraulic cylinders

Chapter 6 | Classification and Types of Hydraulic Actuators

Hydraulic actuators emerge in the market in a wide range of types, sizes, and models to meet different and varied application requirements. Table 6.1 gives a broad classification of hydraulic actuators showing their main types and sub-types.

Table 6.1 Classification of hydraulic cylinders

Main types	Sub-types	Examples
Linear actuators	Single-acting	Spring-returned
		Gravity returned
	Double-acting	Non-cushioned type
		Cushioned type
	Variants	Plunger/Ram cylinder
		Position sensing cylinder
		Double-rod-end cylinders
	Special assemblies	Tandem cylinder
		Telescopic cylinder

Single-acting Hydraulic Cylinders

Figure 6.1 shows the cross-sectional view of a single-acting cylinder. It consists of a barrel, a piston-and-rod assembly, a spring, end-caps, seals, and a port. A fluid chamber is formed in the cylinder with the barrel, piston, and cap-side endplate. The piston-and-rod assembly is a tight fit inside the barrel and is biased by the spring. The seals are primarily used to prevent fluid leakage in the system. The port is integrated into its cap-end for applying or relieving the system fluid. The application of hydraulic pressure through the port moves the piston-and-rod assembly in one direction to provide the working stroke. The piston-and-rod assembly moves in the opposite direction, either by force due to a spring force or by gravity or even by an external force. In a cylinder with a spring-assisted retraction, the spring is designed not to carry any load, but, to retract the piston and piston-rod assembly with sufficient speed. In the alternative design of a single-acting cylinder, the spring can be mounted on the piston side of the cylinder with the port in the head-end. The single-acting cylinder is capable of performing work only in one direction of its motion and hence the name 'single-acting cylinder'.

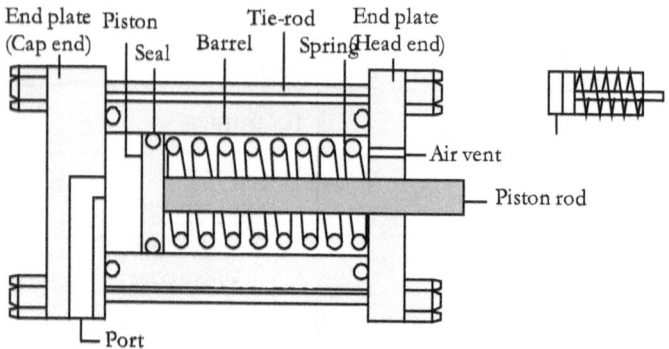

Figure 6.1 | A cross-sectional view of a single-acting cylinder

The stroke length of the cylinder is limited typically to 100 mm due to the natural length of the spring. Single-acting cylinders are linear actuators, which are the simplest in design and economical. A single-acting cylinder converts system pressure and flow into mechanical force and motion, respectively. The ease of operation of the single-acting cylinders makes them particularly suitable for applications, such as clamping, pressing, cutting, holding, ejecting, feeding, and lifting.

Double-acting Hydraulic Cylinders

Figure 6.2 gives the cross-sectional view of a double-acting cylinder. It consists of a barrel, a piston-and-rod assembly, end-caps, seals, and two ports. In contrast to the single-acting cylinder, the double-acting cylinder has fluid ports on both ends, namely the piston-side port and piston-rod-side port. The application of the pressure through the piston-side port extends the cylinder, provided that the pressure from the piston-rod side is relieved.

In the same way, the application of pressure to the piston-rod-side port retracts the cylinder, provided that the pressure from the piston side is relieved. That means the cylinder is capable of converting pressure and flow into mechanical force and motion, respectively. A double-acting cylinder can provide power strokes in both directions of its motion, hence the name 'double-acting cylinder'.

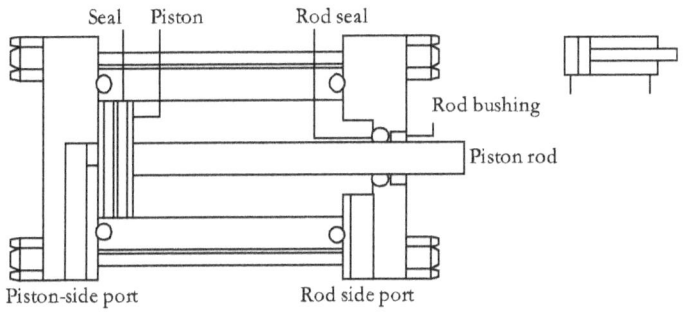

Figure 6.2 | A cross-sectional view of a double-acting cylinder

Ideally, a conventional double-acting hydraulic cylinder can be designed with unlimited stroke length. However, practically the maximum stroke length of the cylinder is limited to about 2000 mm due to the possible bending and buckling of the extended cylinder with a very long piston-rod.

Double-acting hydraulic cylinders are the most common type of cylinders found in modern industries. They can be used in applications, where mechanical power involving linear motion is required. They offer high pushing and pulling forces for the connected loads. They perform well in applications, especially where precise low-speed control is required. They are specially designed for heavy-duty lifting and positioning for mobile applications as well as for industrial production and assembly jobs.

Hydraulic Cylinders - Differential Vs Non-differential

In a typical single-ended double-acting hydraulic cylinder, as shown in Figure 6.3(a), the piston area (say A1) exposed to fluid contact on the piston end is larger than that (say A2) on the piston-rod end. The active area of the fluid contact on the piston-rod end is equal to the bore area minus the cross-sectional area of the piston-rod. Such a cylinder is regarded as a differential cylinder.

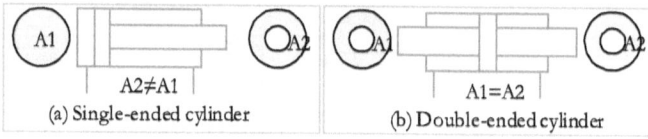

(a) Single-ended cylinder   (b) Double-ended cylinder

Figure 6.3 | Basic types of hydraulic cylinders

The differential cylinder produces somewhat a greater force and lesser speed when extending as compared to the force and speed values when retracting, provided the pressures applied on both ends of the cylinder are the same. The required area ratio (A1/A2) for a differential cylinder depends on its construction, application, and stroke length. For standard applications, the area ratio is typically about 6:5. For heavy-duty applications, with cylinders of large size piston-rods, the area ratio of about 2:1½ may be selected.

In a non-differential hydraulic cylinder, as shown in Figure 6.3(a), the areas of fluid contact on the piston end and the piston-rod end are the same. This fact allows for the generation of equal forces and speeds in both directions of piston travel.

Cushioning in Hydraulic Cylinders

The load-attached fast-moving piston of an ordinary hydraulic cylinder produces impact forces when it strikes on its end covers. These end-of-travel shock loads can be reduced by incorporating cushioning devices in the cylinder either at one end or at both ends. The cushioning devices work by decelerating the piston-

and-rod assembly as it approaches its end-of-stroke position, thus preventing the development of excessive mechanical stresses. The cushions are either fixed type or adjustable type. Hydraulic cylinders can be designed to operate at higher speeds with simple or adjustable cushions.

Hydraulic Cushion Cylinder

Figure 6.4 shows the cross-sectional view of a double-acting hydraulic cylinder with adjustable cushioning devices. A cushioning device consists of a throttle valve integrated into the end caps of the cylinder. Cushion sleeves are attached to the piston.

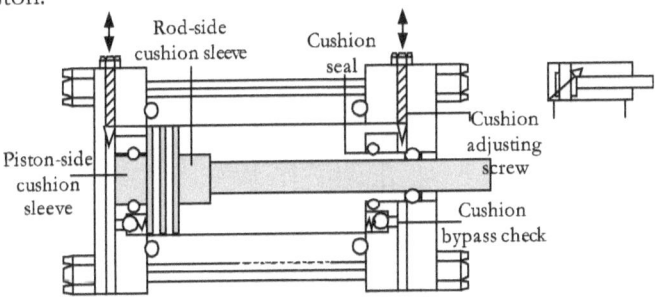

Figure 6.4 | An adjustable cushion hydraulic cylinder

When the piston starts moving forward, full system flow can exit unrestricted through the large orifice in the fluid chamber at the head-end of the cylinder. As the rod-side cushion sleeve enters its chamber, the normal exit path of the system fluid is blocked. This obstruction forces the flow to pass through the throttle valve, which restricts the flow and progressively slows down the piston movement. The piston continues to move to its end-of-stroke position at a controlled speed without causing damage to the cylinder. In many cylinders, the cushioning can be adjusted by using an adjusting screw. Typically, the cap-end cushion would function in the same way as the head-end cushion. The cushions are, usually, designed to operate over the final 20 mm of the piston stroke.

A bypass check valve can be incorporated into the cushioning device to allow the unrestricted entry of the system fluid into the cylinder. Some amount of pressure intensification develops on the piston-rod end but, usually, this is generally not a significant problem. If the load-coupled piston tends to develop enormous forces and high accelerations, extra measures, such as the provision of an external shock absorber, must be taken to assist the load deceleration.

Ram (Plunger) Cylinders

A variant of the standard single-acting hydraulic cylinder is the ram (plunger) cylinder. Figure 6.5 shows the cross-sectional view of the ram cylinder.

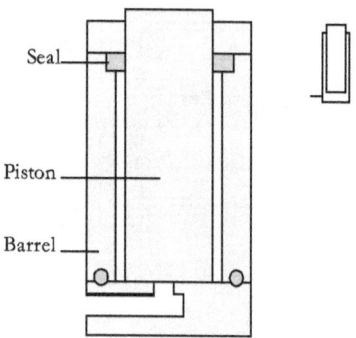

Figure 6.5 | A Plunger/Ram cylinder

A ram cylinder is designed with a robust and thick piston-rod (ram) whose diameter is almost equal to that of the cylinder piston. Most ram cylinders do not use return springs. Instead, they use gravity or the loads attached to them to retract the piston-rods. Both the piston and the piston-rod in a ram cylinder must be precisely cylindrical and finely finished for the smooth output motion. A surface hardened or chrome-plated piston-rod is often used with an ultra-fine surface finish for ensuring the long service life of its sealing elements.

The ram cylinder is, usually, placed in an upright position and is most often used for short-stroke applications. The ram cylinders are primarily used for push operations rather than pull operations. However, a ram cylinder with a hollow rod can be used for pushing as well as pulling operations. The piston-rod bending in a basic long-stroke horizontal cylinder unit or the buckling in the basic cylinder under a high vertical load can be prevented by replacing the basic cylinder with a ram cylinder.

The ram cylinders are most commonly used in high-pressure applications, such as pressing and stamping operations in machines and lifts in automobile service stations.

Double-rod-end Cylinders
A variant of the standard double-acting cylinder is the double-rod-end cylinder. Figure 6.6 shows the double-rod-end cylinder with the piston-rods extending out of the cylinder at both ends.

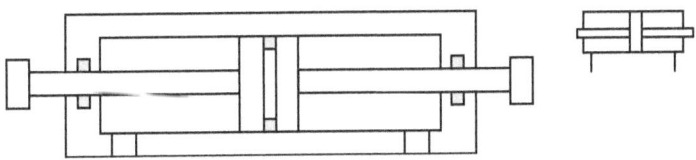

Figure 6.6 | A double-rod-end hydraulic cylinder

The double-rod-end arrangement provides a better rod alignment as the attached piston-rods are moving on two bearings. The cylinder has equal areas on both sides of the piston and operates without the differential cylinder effect. With the use of a double-rod-end cylinder, two functions can be performed at either end in staggered cycles. It is also used when the piston speed must be the same in both directions of its motion. Further, it is useful in a robotic mechanism, where the piston-rods are clamped at both ends of the cylinder, and the body moves instead.

Telescopic Cylinders

A multi-stage telescopic cylinder nests several-cylinder bodies inside one another. That is; the piston-rod of the first stage is used as the barrel of the second stage, and the second piston-rod is used inside the barrel of the second stage. Similarly, there can be up to six stages. Therefore, the total stroke length of the telescopic cylinder can be up to six times the stroke length of the primary cylinder. Remember that the output force is the highest with the first stage and decreases with each new stage.

The telescopic cylinders are ideal for applications that require the use of long-stroke cylinders in a space-constrained environment. They are widely used in hydraulic equipment for the agriculture, construction, and heavy engineering industry. They are commonly used in mobile hydraulic systems for the tilting of truck dump bodies and forklifts, lifting operations in hydraulic cranes, and material handling. The telescopic cylinders are constructed of single-acting and double-acting varieties.

Single-acting Telescopic Cylinder

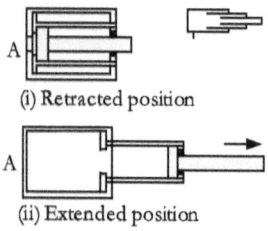

(i) Retracted position

(ii) Extended position

Figure 6.7 | A single-acting telescopic hydraulic cylinder

Figure 6.7 shows the single-acting telescopic cylinder. It is a multi-stage cylinder with two to six concentric tubular envelopes. Most telescopic cylinders are of the single-acting type, where the fluid pressure always acts in one direction. That is; on the forward strokes, to be more precise.

Double-acting Telescopic Cylinder

Figure 6.8 shows the double-acting telescopic cylinder. In this type of cylinder, the system pressure acts alternately for extending and retracting the cylinder. The double-acting telescopic cylinders are highly complicated and must be specially designed and manufactured to a high degree of precision. Therefore, they are much more expensive than conventional hydraulic cylinders.

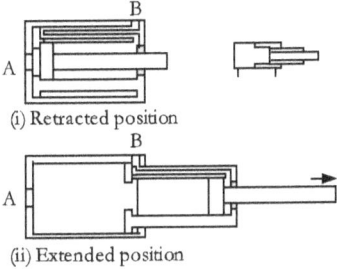

(i) Retracted position

(ii) Extended position

Figure 6.8 | A double-acting telescopic hydraulic cylinder

Tandem Cylinder

In the tandem type of cylinder, two (or more) cylinders are assembled in-line, with the piston-rod of the first cylinder attached to the piston of the second cylinder (and so on). Figure 6.9 shows the cross-sectional view of the two-stage tandem cylinder.

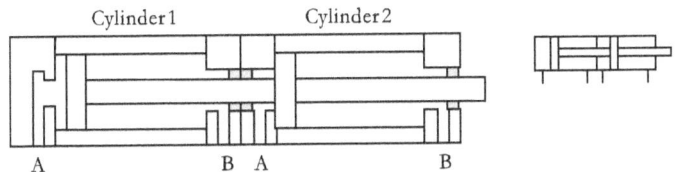

Figure 6.9 | A cross-sectional view of a tandem cylinder

When pressurised fluid is applied to the piston-side ports (marked as 'A'), the cylinder extends with force, which is twice as much as the force produced by a single-stage cylinder of the same bore size while extending. In the same way, when pressurised fluid is applied to the piston-rod-side ports (marked as B), the cylinder retracts with force, which is twice as much as the force produced by a single-stage cylinder of the same piston and piston-rod sizes while retracting. The tandem cylinder is suitable for an application where a high output force must be developed within a narrow radial space, but there is enough axial length.

Swing Clamp Hydraulic Cylinder

It is a cylinder with a clamp arm. It can swing and clamp. The piston and piston-rod assembly of the cylinder can rotate by a certain angle in the clockwise or anti-clockwise direction during the swing stroke and then travels in a straight line during the clamp stroke. It is meant for the secure and safe clamping and unclamping of work-pieces without obstruction. However, the arm should not contact the work-piece during the swing stroke. Swing cylinders are designed in a single-acting version with an integrated return spring or double-acting version. In applications where return time is critical, a double-acting cylinder can ensure a positive retraction on a timely basis.

Construction of Swing Cylinder

Figure 6.10 shows the cross-sectional views of a swing cylinder in the extended and retracted positions. It consists of a barrel, piston, and piston-rod with an arm or lever. The piston is a tight-fit into the barrel. The arm of a standard length is attached to the piston-rod on one side of the piston for the clamping operation. The piston-rod on the other side of the piston has ball seats that have straight and angled sections along their axis, as shown in the Figure. Cam followers made of tungsten-carbide, contact and guide through the ball seat on the piston-rod. This sturdy swing mechanism controls the swing and straight motions of the piston and piston-rod assembly during its extension and retraction strokes.

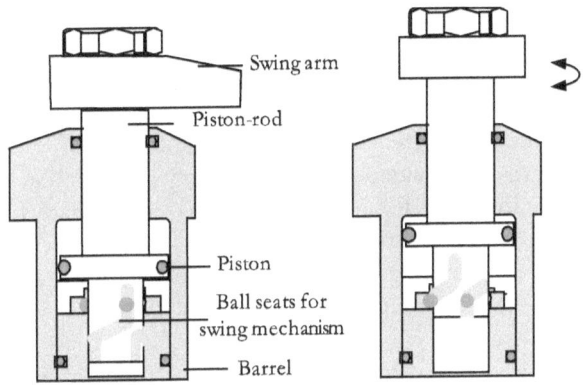

Figure 6.10 | A swing clamp cylinder in the extended and
retracted positions

Working Principle of a Swing Cylinder

The swing cylinder clamps a work-piece while retracting and unclamps while extending. The movement of the piston and piston-rod assembly with an arm consists of a swing stroke and clamp (or unclamp) stroke. The piston-and-piston-rod assembly with the arm turns during the swing stroke. It then retracts in a straight line to apply clamping force, with the arm in position over the work-piece during the clamping stroke. It extends in a straight line as it releases the work-piece during the unclamp stroke and then the piston along with the arm rotates to the full extended position during the swing stroke, providing easy access for loading and unloading of the work-pieces.

Typical Specifications, Swing Cylinders

Table 6.1 | Specifications of swing cylinders

Bore size	14, 25, 32, 40, 50, 63, 80 mm
Clamping stroke	6, 7, 8, 10, 11, 13, 14, 15, 22, 25, 50 mm
Swing angle	0°, 30°, 45°, 60°, 90°
Maximum operating pressures	500 bar

Chapter 7 | Position Transducers for Hydraulic Cylinders

As the demand for greater functionality and ease of control of hydraulic cylinders increases, as in the heavy industry and mobile equipment, the need for hydraulic cylinders fitted with position transducers (sensors) is becoming more critical. A position transducer is essentially a feedback device used with a cylinder that senses the position of the cylinder's piston rod.

A position transducer typically consists of two parts: one part is fixed to the cylinder body, while the other part moves with the piston-rod whose position is being measured. It can be mounted externally or integrally to the piston-rod. If a transducer is to be mounted inside a cylinder, a hole must be drilled through the piston rod. Further, the cylinder end cap is machined to accommodate the transducer. The initial cost of gun-drilling the rod and the replacement cost can be the limiting factor in the usage of integrally mounted transducers.

Three types of position transducers are typically used in hydraulic cylinders. They are: (1) Variable resistance potentiometers, (2) Linear Variable Differential Transducers (LVDTs), and (3) Magnetostrictive transducers.

Linear Resistance Potentiometers

A linear resistive transducer is a linear potentiometer used to detect the position of the piston-rod. It can be an external cable connected to a rotary potentiometer or an internally mounted flat potentiometer contained within the piston-rod. The piston-rod supports an electrically conductive wiper running along the surface of a partially conductive plastic probe. As the wiper moves over the plastic element, its resistance changes in a linear fashion, making it easy to determine the position of the piston-rod. Linear resistive transducers are low-cost devices, but they have low accuracy.

Linear Variable Differential Transducers (LVDTs)

An LVDT is a non-contact transducer that converts linear displacement into an electrical output signal. It consists of primary and secondary coils. This coil assembly is then installed in the piston of the cylinder. A ferromagnetic core moves inside the coils. An alternating current in the primary winding creates an axial magnetic flux. This flux is coupled to the secondary windings through the core, inducing an output voltage in each secondary winding. The series-connected secondary windings produce a net voltage in relation to the position of the core with the help of built-in electronics. LVDTs have the highest accuracy and can withstand shocks and vibrations in the associated systems.

Magnetostrictive technology

Magnetostriction is a property of ferromagnetic materials that causes them to change their shape or dimensions in the presence of a magnetic field. A magnetostrictive linear position transducer, as shown in Figure 7.1, uses an iron-alloy sensing element, typically called a waveguide. The waveguide is housed inside a pressure-rated stainless steel tube. Further, a conductor and signal converter are attached to the waveguide. The piston of a hydraulic cylinder is fixed with a permanent magnet.

Initially, a short-duration electrical pulse (1 – 3 μsec) is applied to the conductor. The current creates a magnetic field along the waveguide. The magnetic field then interacts with the generated magnetic field from the position magnet, inducing a torsional deflection of the waveguide element. When the current pulse stops, the strain is abruptly relaxed, causing a mechanical pulse to be propagated along the waveguide. This mechanical pulse travels in both directions at a constant speed (2850 metres/second) and is detected at the signal converter. The detection of the mechanical wave in the signal converter completes one measurement cycle.

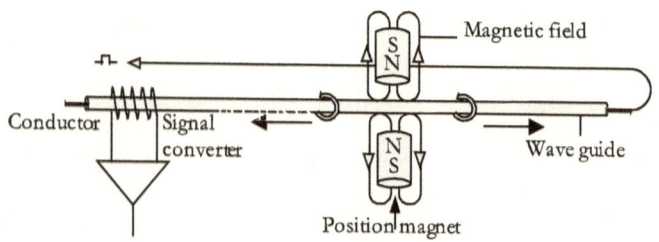

Figure 7.1 | A magnetostrictive transducer

The time between the initial electrical pulse and the received mechanical pulse accurately represents the absolute position of the magnet and, hence, the position of the hydraulic cylinder. The position of the magnet along the waveguide is calculated by accurately timing the interval between the initial current pulse and the detection of the mechanical return pulse. Measurement cycles are typically repeated at rates of 0.5 to 5 milliseconds, depending on the length of the sensor.

The linear position sensors are the ideal choice for hydraulic cylinder position feedback. They are available in measuring lengths from 50 mm to 7620 mm and can withstand high-pressure operation up to 600 bar. Many electrical interface options are available to adapt to any control system. They are exceptionally well-suited for use in hydraulic cylinders used in industries such as steel processing and tyre manufacturing.

Chapter 8 | Installation and Mounting of Hydraulic Cylinders

The body of a hydraulic cylinder is to be mounted properly, and its piston-rod is to be coupled correctly to the machine for the optimum performance of the cylinder. Any mismatch in the mounting or coupling can result in the development of side loads. Remember, the side loads act laterally across the barrel, piston, piston-rod, seals, and bearing bushes. The side loads will cause stresses on the cylinder parts leading to the lowering of service life of the cylinder.

A hydraulic cylinder should be so fixed that the side loads on the piston-rod bearing must be a minimum. By taking advantage of good engineering practices, they can be reduced to an acceptable limit. Slide or roller guides can be used to carry the load, wherever possible. The mounting dimensions of hydraulic cylinders must meet the requirements of the relevant standards under the ISO/NFPA.

Mounting Methods of Hydraulic Cylinders

A cylinder can be mounted to a machine with a permanent type or a movable type of mounting. For example, the body and the piston-rod of a hydraulic cylinder can be rigidly fastened to the machine as in the fixed type mounting, or the cylinder can be allowed to swivel as part of the linkage in one or more planes as in the pivot type mounting. Figure 8.1 shows the schematic diagrams of these two configurations.

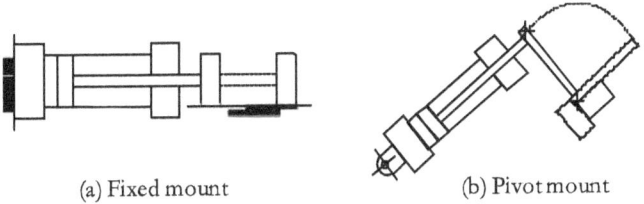

(a) Fixed mount   (b) Pivot mount

Figure 8.1 | Mounting arrangements for hydraulic cylinders

The selection of a mounting method for a hydraulic cylinder depends primarily on the characteristics of the given application. Factors, such as the cylinder stroke length, piston-rod diameter, method of connection to the load, and presence of shock pressures must be considered when selecting the mounting style for the cylinder.

Mounting Styles of Hydraulic Cylinders
Cylinder mounts and accessories are available in various configurations. The mounts can be fixed type or movable type. The typical examples of the fixed mountings are the tie-rod mounts, flange mounts, and foot mounts, and the examples of the pivot mounts are the head, cap, or centre trunnions, swivel flanges and rear cap pivot mounts. The load-bearing capacity of a cylinder mount should exceed the load of the associated cylinder. Some critical mounting methods for the cylinder body are depicted in Figure 8.2(a) and some coupling methods for the piston-rod are depicted in Figure 8.2(b).

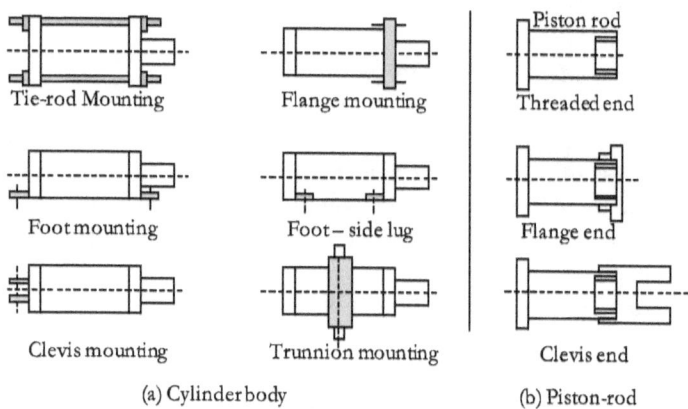

Figure 8.2 | Typical mounting arrangements of cylinders

Tie-rod mount: In this type, the extended tie-rods of a hydraulic cylinder are used to mount the cylinder. The tie-rods can be extended either on the piston side or the piston-rod side. The free end of the cylinder should be supported to prevent its misalignment or sag. The best application of the tie-rods extended on the piston side is with a thrust load (i.e., with the piston-rod in compression), and that of the tie-rods extended on the piston-rod side is with a tension load.

Flange Mount: This type of mount provides a sturdy and rigid mounting for a cylinder. With this type of mount, there is no scope for misalignment. The flange mounting is the best for installing the cylinder when a load force acts along the centerline of the cylinder. The best use of a cap-end flange in the cylinder is in a thrust load application. The piston-rod-end flange mounts are best used for tension load applications.

Foot or Lug Mount: The foot-mounted or lug-mounted cylinder provides relatively rigid mounting. This type of cylinder mounting can tolerate some amount of misalignment.

Pin-and-Trunnion Mounts: A pin-and-trunnion-mounted hydraulic cylinder needs a provision for pivoting on both ends of the cylinder. It is designed to carry shear loads. If the path of the load is curved or misalignment is a problem, the pivoted centerline mounting should be used.

Piston-rod Mounts: Various forms of coupling accessories used for the piston-rods of cylinders are the pinhole, rod-end threads, rod clevis, eye bracket, knuckle, and pivot pin. Figure 8.2(b) shows some coupling arrangements for the piston-rods.

Threads: The most important threads used are the SI system Thread, British Standard Pipe Thread, National Pipe Thread, and Unified Fine Thread. Essential specifications for the piston rod end include its thread diameter and pitch and the thread length.

Chapter 9 | Advantages of Hydraulic Cylinders

Some of the significant advantages of hydraulic cylinders are their ruggedness, power density, and economy. They are also available in a broad range of strokes. The following section explains these positive features:

Ruggedness: Hydraulic cylinders are generally constructed as robust actuators. If a hydraulic cylinder is highly overloaded, it just stops. It does not overheat or burnout.

Power Density: A hydraulic cylinder can produce an enormous amount of force for its comparatively small size. This powerful actuator can easily be built to fit the tight confines of modern machinery.

Range of Strokes: Small industrial hydraulic cylinders, such as clamping cylinders, may have a maximum stroke of 6 mm. Large hydraulic cylinders, as used in the heavy equipment industry, can have power strokes up to 2 m. Very long stroke lengths may be obtained by using telescopic cylinders.

Economy: Hydraulic cylinders are relatively easy to manufacture to the exact dimensions and speed requirements, as specified by the end-users.

Reliability: Hydraulic cylinders are highly reliable, even in a harsh industrial environment.

Chapter 10 | Applications Notes, Hydraulic Cylinders

Manufacturers offer a range of standard cylinders and heavy-duty cylinders suitable for a variety of applications and harsh environments. Hydraulic cylinders can range from single-acting or double-acting to telescopic or plunger type cylinders. The heavy-duty hydraulic cylinders are designed for high-pressure, high-flow, high-force applications, and corrosive environments. They are used in all kinds of industrial and mobile applications for obtaining the pulling and pushing operations, especially with the precise low-speed control requirements. The following list highlights some of the application areas:

- Steel mills
- Defence sector
- Automotive plants
- Construction sector
- Chemical and food industries
- Industrial production and assembly jobs
- Mining, offshore, and aerospace applications
- Metal-cutting, machine tools, and forge-pressing machines
- Clamping, pressing, cutting, holding, feeding, lifting, and material handling operations
- Mobile applications for heavy-duty digging, hoisting, tilting, and positioning operations
- Hydraulic cylinders are used in transportation equipment, hoists, and are required for many nautical steering mechanisms in shipbuilding

The growing applications of hydraulic cylinders in industrial machines, agriculture machinery, and material handling equipment in infrastructure development, will raise the demand for the cylinders.

The self-explanatory Figure 10.1 gives some typical examples of hydraulically-driven mechanisms.

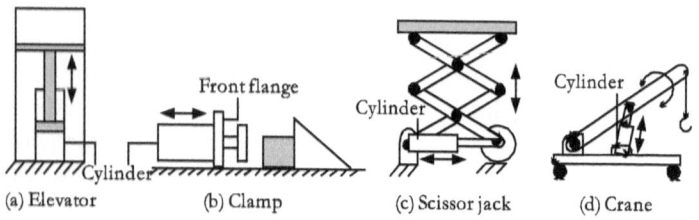

Figure 10.1 | Examples of hydraulically-driven mechanisms

Applications of Swing Cylinders

Hydraulic swing clamps are used for clamping of work-pieces when it is essential to keep the clamping area free of fastening straps and clamping components for the unrestricted loading and unloading of the work-piece. A typical application of swing cylinders is shown in self-explanatory Figure 10.2.

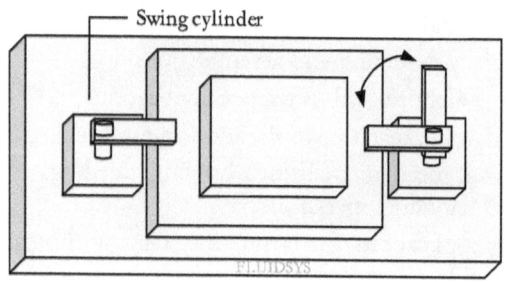

Figure 10.2 | An example of work-piece clamping

Note: Appendix 3 gives important specifications to be considered while selecting hydraulic cylinders.

Chapter 11 | Standards, Hydraulic Cylinders

Hydraulic cylinders are available conforming to the ISO and NFPA standards. The ISO hydraulic cylinders are manufactured conforming to the standards: 6020-1, 6020-2, 6020-3, and 6022. The ISO hydraulic cylinders have standard installation dimensions, and hence they are interchangeable. Table 11.1 gives some ISO standards for hydraulic cylinders.

Table 11.1 | Important ISO standards for hydraulic cylinders

ISO 6020-1:2007	This standard establishes metric mounting dimensions for medium series cylinders with bores from 25 mm to 200 mm for pressures up to 160 bar. The medium series dimensions apply to round head cylinders.
ISO 6020-2:2006	This standard establishes metric mounting dimensions for compact series cylinders with bores from 25 mm to 200 mm for pressures up to 160 bar. The compact series dimensions are most applicable to square head cylinders.
ISO 6020-3:2006	This standard establishes metric mounting dimensions for compact series cylinders with bores from 250 mm to 500 mm for pressures up to 160 bar. The compact series dimensions are most applicable to square head cylinders.
ISO 6022:2006	Establishes mounting dimensions for hydraulic cylinders for use at 250 bar
ISO 3320:1987	This standard establishes a metric series of bores and diameters fluid power cylinders.
ISO 4395:2009	This standard specifies thread dimensions and configurations for use with hydraulic cylinder piston-rod-ends.
ISO 6099:2009	This standard specifies a code for identifying cylinder mounting, envelope, accessory and connector dimensions, and for cylinder mounting and accessory types.

Chapter 12 | Maintenance and Safety of Hydraulic Cylinders

Maintenance of Hydraulic Cylinders

When installed and operated correctly according to the manufacturer's instructions, hydraulic cylinders are generally maintenance-free. The most critical elements of a hydraulic cylinder, from the maintenance point of view, are its seals and piston-rods. The surface of the piston-rod that enters the piston-rod gland must be kept smooth and clean to avoid the failure of piston-rod seals and glands and prevent the wearing of cylinder bearings. Impact forces can also damage the seals in the cylinder.

Other issues of concern are the presence of air, internal and external fluid leakages, and its operation in harsh environmental conditions. Specialised components, such as metal rod scrapers or protective piston-rod boots, can be provided with the cylinder in harsh and dirty environments for protecting the gland area of the cylinder piston-rod. The presence of air contamination or the occurrence of fluid leakage or maintaining an extremely viscous fluid medium in the cylinder may cause the sluggish or erratic operation of the cylinder. The contaminated fluid inevitably leads to an early failure of the cylinder.

Further, the cylinder must be aligned and mounted correctly with its associated load part for reducing the side load on the cylinder and avoiding the consequent failure of cylinder seals, minimising the wear of cylinder bushing, and preventing the bending or breakage of the cylinder piston-rod. Ensure that appropriate equipment is available to handle the repair activities safely on the cylinder if it is heavy. The following sections explain the various aspects of cylinder maintenance.

Essential Maintenance Activities for Hydraulic Cylinders

Apart from the general maintenance activities, the following maintenance activities can be carried out on the cylinder to keep it in good working condition:

- Check the piston-rod for straightness. Check for any dents or damages on the piston-rod due to impact forces
- Examine the piston-rod bearing for roundness
- Examine the barrel, the piston and the piston-rod for nicks, scoring, and pitting
- Check the cylinder for worn components
- Check and control the internal and external fluid leakages in the cylinder
- Replace piston seals, piston-rod seals and/or piston-rod bushings, if leakages occur
- Align the cylinder and its mating part in line, to avoid side loads on the cylinder
- Check the cylinder mountings periodically for tightness or cracks
- Check for sluggish/erratic operation of the cylinder
- Check for the creeping of the cylinder
- Open the bleed ports provided in the cylinder for releasing the trapped air in the cylinder
- After an initial operating period of typically 40 hours, the screws of the cylinder head and bottom and all fixing screws should be retightened to the torque specified
- Ensure lubrication of the bearing points, such as the swivel and articulated bearings as well as the swivel journal

Cylinder Faults

Table 12.1 | General faults

Damaged/worn seals	-Replace seals
Excessive rod wear	-Prevent wear
Rod damaged	-Replace rod
Rod seized	-Repair
Faulty alignment	-Align properly
Broken linkages	-Replace damaged linkages
Loose mountings	-Tighten mounting bolts
Jerky movement	-Assemble piston packing correctly -Avoid heavy load at slow speed
No thrust	-Replace faulty piston -Prevent excessive leakage -Set PRV to correct pressure
Abnormal cylinder thrust	-Bleed entrained air from the cylinder -Check and set cylinder cushion

Table 12.2 | Erratic operation

Air in system	-Replace worn seals -Drain and flush fluid -Fill the reservoir to the proper level -Tighten leaking connections -Bleed air -Replace worn shaft/seal
Cylinder/valve sticking/binding	-Remove dirt/gummy deposit -Remove fluid contamination -Remove worn parts -Install seals properly
Cylinder internal leakage	- Repair or replace worn parts -Tighten loose packing -Use medium viscosity fluid -Control fluid contamination -Control wear

Safety Instructions

Many precautions have to be taken while installing, handling or maintaining hydraulic cylinders. Some of the essential precautions are highlighted below. Failure to follow the precautions may result in personal injury or property damage or both.

This is the safety alert symbol. It is used on labels and instruction manuals of a machine to alert the users to potential personal injury hazards. Therefore, it is essential to obey all safety messages that follow this symbol to avoid possible injury or death.

- Depressurise a cylinder before repairs are made.
- Any risks of injury should be avoided while installing and testing hydraulic cylinders.
- Read the owner's manual completely and familiarise yourself thoroughly with the product and its components and recognise the hazards associated with its use.
- The owner and operators shall have an understanding of the safe operating procedure of cylinders and the system before attempting to use them.
- The operator must be trained, qualified, and familiar with the correct operation, maintenance, and use of cylinders and the associated machine.
- Instructions and safety information shall be given in the operator's native language before the use of the product is authorised.
- Make sure that the operator thoroughly understands the inherent dangers associated with the use and misuse of the product.
- Wear personal protective equipment (PPE) when installing, operating or maintaining hydraulic equipment.

- Stay clear of a lifted load before it is adequately supported.
- A hydraulic cylinder, when used as a load lifting device, should never be used as a load-holding device. After the load has been raised, it should be blocked
- Never attempt to lift a load weighing more than the capacity of the cylinder. Overloading causes equipment failure and possible personal injury
- Never rely on hydraulic pressure alone to support a load
- Keep hands and feet away from a cylinder and work-piece during the operation of the cylinder
- Do not subject a cylinder to shock loads
- A cylinder must be able to support the load while pushing or lifting and hence it must be on a stable base
- Ensure cylinder is fully engaged into adapters and extension accessories
- A cylinder should not be subjected to off-centre loads to avoid side loads
- Entrust the repair of hydraulic cylinders to qualified and authorised personnel
- All cylinder fittings should be tightened with proper tools for leak-free connections
- Do not over-tighten the connections. Over-tightening can cause premature thread failure
- Pay attention to the possibility of over-pressurisation in cylinder chambers. If required, install additional pressure relief valves or reduce the working pressure
- If larger masses are to be stopped by end-cushion devices, make sure that the cushioning pressure is not too high. The final adjustment must be made on the control side to obtain the most effective cushioning
- The load applied to the piston of a long-stroke cylinder should not exceed the permissible force to avoid the buckling of the piston-rod
- Keep hydraulic cylinders away from flames and heat

Chapter 13 | Design of Hydraulic Cylinders

Materials for Cylinder Barrels

The most common material for hydraulic cylinder barrels is the low carbon steel or low carbon cold-drawn seamless steel tubing, with a tensile yield point of about 4000 bar. This material can be used to construct cylinders rated up to 600 bar and 125-mm bore. To construct cylinders of higher pressure rating or bore size, ductile steel with a higher yield point should be used. Cast iron should never be used for pressures over 140 bar regardless of wall thickness.

Burst Pressure

The wall thickness of steel tubing is considered thin, if it is less than 10% of the tubing ID, and the wall thickness is considered thick if it is greater than 10% of the tubing ID. Barlow's formula can be used to find the burst pressure of a thin-wall steel tubing and is given by:

$$\text{Burst pressure (BP)} = 2tS/ID$$

Where,

t = Wall thickness, mm
S = Tensile strength of the tubing material, bar
ID = Inside diameter, mm

The burst pressure can be calculated by the following formula:

$$\text{Burst pressure (BP)} = S \times (R^2 - r^2) / (R^2 + r^2)$$

Where,

S = tensile strength of the tubing material, bar
R = Outside radius, mm
r = Inside radius, mm

Working Pressure

The working pressure of a conductor is the safe pressure to which it can be subjected. It is calculated by dividing the burst pressure of the conductor by a safety factor.

$$\text{Working pressure (WP)} = \frac{\text{Burst pressure (BP)}}{\text{Safety factor (SF)}}$$

- A safety factor of 4:1 is used for hydraulic applications where shock and mechanical strain are not considerable.
- A safety factor of 6:1 should be used where considerable shock and mechanical strain are expected.
- A safety factor of 8:1 should be used where severe hydraulic shock and mechanical strain are expected.

Example 13.1

A hydraulic cylinder made of mild steel has an inside diameter of 150 mm and a wall thickness of 20 mm. What is the maximum working pressure that can be applied to the cylinder, if the tensile strength of the material of cylinder construction is 3500 bar? Assume a safety factor of 5.

Solution

Inside diameter, ID	= 150 mm
Wall thickness, t	= 20 mm
Outside diameter, OD	= 150 + 2x20 = 190 mm
Tensile strength, S	= 3500 bar
Safety factor	= 5
Inside radius, r	= 150/2 = 75 mm
Outside radius, R	= 190/2 = 95 mm
Burst pressure	= S x $(R^2 - r^2)/(R^2 + r^2)$ [t >10% of ID]
	=3500 x$(95^2 - 75^2)/(95^2 + 75^2)$
	=3500 x (9025–5625)/(9025+5625)
	=3500 x 3400 /14650 = 812 bar
Working pressure	= Burst pressure / Safety factor
	= 812/5 = 162 bar

Chapter 14 | Manufacturing Process of Hydraulic Cylinders – Salient Points

Introduction
Hydraulic cylinders are used in many types of machines and systems for industrial and mobile applications. As we are aware, they operate through high-pressure fluid media. Further, they must withstand high temperatures and high stresses present in applications.

The manufacturing process of cylinders involves various steps including heat treatment and machining. The manufacturing process can use in-house machinery such as grinding, boring, drilling, honing, and welding machines. The process may also use many other auxiliary systems for heat treatment, testing, etc. Alternatively, many work processes can be outsourced for reasons of economy. But remember, outsourcing may impair the quality of cylinders being manufactured.

The manufacturing of high-quality cylinders, aiming for good surface quality and geometrical accuracy, requires extensive engineering and design capabilities, the selection of high-quality materials of construction, high-performance manufacturing systems, and expert workmanship. The following sections present the salient points of the manufacturing process of hydraulic cylinders for an initial understanding.

Steps for Cylinder Manufacturing
The manufacturing process of hydraulic cylinders mainly involves the following steps: design, heat treatment, machining, coating, assembly, and testing.

The Design Phase, Cylinder Manufacturing
The design phase of a cylinder is an important stage in the process of cylinder manufacturing. The design must consider its operating conditions, performance requirements, size constraints, and cost limitations.

In the design stage, the requirement specifications of the cylinder to be designed are transformed into technical drawings describing the materials of construction, dimensions, tolerances, quality of internal surfaces, and coating methods. The design can take advantage of various software packages for quickly and accurately preparing drawings.

Parts of a Hydraulic Cylinder

The manufacturing process of a hydraulic cylinder mainly involves the preparation of its constituent parts and their assembly. A hydraulic cylinder usually consists of the following: barrel, piston, piston rod, end plates, sealing (piston seals, rod seals, wear bands, wiper seal etc), ports, and mounting accessories.

Parameters of Hydraulic Cylinders

The major parameters of a hydraulic cylinder may include the following:
-Inner diameter (Bore diameter)
-Wall thickness
-Piston outer diameter
-Piston-rod diameter
-Stroke length
-Depths of seal grooves
-Widths of seal grooves
-Sizes of the seal ring holes
-Roundness
-Plating thickness

Machinery and Equipment for Cylinder Manufacturing

High-capacity CNC lathes, milling machines, boring machines, drilling machines, and welding machines are required for the efficient machining of the parts. Highly specialized CNC machines can give flexibility to machine parts of complex and non-standard hydraulic cylinders. Other types of machinery or equipment are also required for operations such as coating, assembly, and testing.

Barrel Preparation
A barrel is the main part of a hydraulic cylinder. The barrel along with the piston, and end covers should form a sealed chamber for containing the pressurized system fluid. The barrel should be inseparably welded to the cylinder mounting fixtures.

Requirements of Barrel
The barrel should have sufficient strength to prevent deformation when subjected to operating as well as shock pressures. It should also have enough rigidity to prevent bending when subjected to lateral forces.

The barrel should also have smooth interior surfaces with a small roughness value, high precision tolerances, and a durable service life.

Further, the barrel that is required to be welded to a flange, mounting fixtures, or joints should have good weldability to avoid cracking or excessive deformation of the parts after welding.

The requirement specifications for barrels in hydraulic cylinders can be realised by a proper selection of barrel material. The barrel materials should have high strength, high resistance to wear and corrosion, good machinability, good weldability, and low cost.

Heat Treatment
Heat treatment is an essential process during the manufacturing process of hydraulic cylinders. It is the process of heating a metal part to a specified temperature, holding (soaking) it at that temperature for a specified duration (soak time), and then cooling it back, at a controlled rate, using specific methods to get desired properties.

During the heat treatment process, the metal part will change its physical structure and mechanical properties such as hardness, ductility, brittleness, and corrosion resistance. Some of the most common heat treatment methods include annealing, hardening, tempering, carburisation, etc.

In the **annealing process**, the metal is heated beyond its upper critical temperature and then cooled at a slow rate. Annealing is carried out to soften the metal to improve its machinability, ductility, and toughness.

In the **hardening process**, a metal part is heated to a specified temperature, then cooled rapidly (quenching process) by submerging it into a cooling medium such as oil, brine, or water. This process is carried out to increase the hardness of the part. However, the part can become more brittle.

In the **tempering process**, the temperatures are usually set much lower than hardening temperatures. This process is carried out to reduce excess hardness, and therefore brittleness, induced during the hardening process.

In the **carburisation process**, the metal part is heated in the presence of another material, such as carbon monoxide, that releases carbon on decomposition. The released carbon is absorbed into the surface of the metal part. The surface of the metal part gets hardened as the carbon content of the surface increases.

Types of Stainless Steel

Stainless steel is an alloy of steel that contains chromium, nickel, molybdenum, and/or titanium. The addition of elements to steel improves its properties such as hardness, ductility, corrosion resistance etc. For example, the main role of Cr in steel is to improve the properties such as their strength, hardness, and impact toughness after quenching and tempering processes.

Stainless steel can be classified into one of five types. They are austenitic, ferritic, martensitic, duplex, and precipitation. Each of these types can again be subdivided into grades of stainless steel.

The stainless steel families are specified by the ratio of the metals that compose the alloy. An example of a commonly used grading system comes from the American Iron and Steel Institute (AISI). Each category is further divided into series and grades. The grades reflect the specific alloy's durability, quality, and temperature resistance.

The high levels of chromium in austenitic steels make them a good choice for corrosion resistance. Stainless steel alloys containing chromium and nickel often provide the best combination of strength, ductility, and toughness.

Materials for Barrel
There are many steel materials available in terms of their tensile strengths. A barrel for a hydraulic cylinder can generally be prepared from an annealed cold-drawn or hot-rolled seamless steel tube or it can be forged. The steel type is usually carbon steel. However, in applications with corrosive environments, stainless steel can be used. The barrel with a thick wall should be made with cast iron or forgings.

The commonly used materials include:
-Carbon steel: #20 steel, #35 steel, and #45 steel, St 52
-Low tensile carbon steel: AISI 1020
-Medium tensile carbon steel: AISI 1045
-Ordinary structural low-alloy steel: 15MnV and 27SiMn
-Structural alloy steel: 30CrMo, 35CrMo, and 35CrMo-ALA
-Stainless steel: Cr18Ni9
-Low alloy steel: AISI 4140
-Austenitic steel: AISI 304 (18 Cr/8 Ni), AISI 316
-Ferritic steel: AISI 409, 430, 439, and 441
-Duplex grade steel: 2205 (22%Cr + 5%Ni)
-Cast steel: ZG230-450 and ZG310-500

Inspection of Materials for Barrel

It is essential to inspect the tube materials for any defects. Test certificates issued by manufacturers for their chemical composition and mechanical properties must be verified for their suitability.

Rough Turning and Other Operations, Barrel Preparation

First, the cut off the correct length of the tube in a parting operation. Then, turn the outer surface of the barrel to reduce the exterior of the barrel near to the required diameter and shape. Further, carry out chamfering operations to remove the sharp edges in the barrel and add the required external threads.

Machining Process (Boring) for Barrel

Machining is a metal removal process that depends on boring, cutting and grinding operations to remove unwanted material from the barrel to achieve a final shape. Before boring operation, place the barrel in the holder of the boring machine and fix it. Use bolts to tighten and adjust the height of the boring tooltip so that it is consistent with the centre of the barrel.

Carry out the boring operation to finish the semi-finished barrel's inner surface. The boring operation is the main process of machining the barrel. It is essentially a turning operation performed on the internal surfaces of the barrel for achieving dimensional and surface finish tolerances and hence, for realizing a high dimensional accuracy.

Precision Machining, Barrel

After cold drawing and heat treatment, the partly finished barrel is additionally prepared for improving the surface finish and the geometric form of the inner surface of the barrel using the following methods: (1) Honing process and (2) SRB (Skiving & Roller Burnishing) process.

Honing Process

In the honing process, the inside surface of a barrel is ground using abrasive stones and paper. The polishing stone turns while being moved in and out of the barrel. This process can take away small amounts of material from the inner surface, remove surface imperfections, and achieve the precise dimensions tolerances of the inside diameter to provide the best lubrication properties and low wear on cylinders and seals.

It may be noted that ID dimensional tolerances as precise as 0.021mm (H7) and surface roughness in the range Ra 0.2 to 0.8 um can be achieved for the barrel using the honing process.

Skiving and Roller Burnishing (SRB) Process

The skiving and roller burnishing (SRB) process uses a skiving tool on the forward stroke to skive the inside diameter of a barrel to the required size. It uses roller bearings on the return stroke to burnish the surface to the required roughness. As compared to the honing process, the SRB process is faster and can produce more precision tolerances and a smoother surface finish.

Honed steel tube manufactured with the SRB process typically has ID dimension tolerance as precise as 0.021mm (H7) and inside surface roughness smoother than Ra 0.4 um.

Precision Turning, Barrel Preparation

Turn the outer circle of the barrel to achieve the correct size and shape for the barrel. Drill the fluid ports, if the ports are designed to be on the barrel.

Welding, Barrel Preparation

Weld pipe joint seats, flange, and other accessories.

Final stages of Barrel Preparation

During the final stages of barrel preparation, it is brushed with oil for anti-rust treatment. Then, it is moved to a location with all precautions for safe storage.

Preparing the Piston

The piston is first machined for correct sizing. It must then be quenched and tempered for hardness. Verify its hardness through a hardness test.

The correctly-sized piston is, then, machined with grooves to fit seals and bearing elements. The piston is inseparably attached to the piston rod using threads, bolts, or nuts. The piston must be coated with non-ferrous metals for corrosion resistance and precise guidance.

Preparing the Piston-rod

The piston-rod is typically made of steel or stainless steel. Check the material certificate furnished by the manufacturer for its suitability for the piston-rod.

The piston-rod must be machined for obtaining the correct shape and size. First, it must be roughly machined. It must then be quenched and tempered for hardness. Verify its hardness through a hardness test.

It must then be heat treated through induction hardening, carburizing, or nitriding method to improve its surface hardness.

Induction hardening can improve the mechanical properties of the piston-rod, resist buckling failure in a push operation, and protect it from damage in the event of external impact.

It must then be precision machined, finely ground, and polished to provide a reliable seal and prevent leakage. The rod ends must be finished with rounded edges.

The piston-rod is then coated with hard chrome plating or subjected to surface heat treatment. It must then be polished to provide a reliable seal and prevent leakage. The piston-rod needs to be highly resistant to pitting, corrosion, and wear.

Check the prepared piston-rod for hardness and dimensional accuracy. Coat the piston-rod with oil for rust prevention and storage.

Preparing the Piston Guide Sleeve
Check the certificate for the material of the piston guide sleeve for its suitability. Carry out rough machining followed by precise machining for the correct sizing of the sleeve. Inspect the prepared sleeve. Coat the sleeve with oil for rust prevention and storage.

Preparing the Cylinder Head
The manufacturing process of a cylinder head with rod sealing, guiding, cushion, and porting arrangements, involves the inspection of the material certificate, rough machining, quenching and tempering, hardness testing, precision machining, internal and external threading, sawing, milling, boring and fitting, and final inspection.

The head is connected to the body using threading, bolts, or tie rods. An O-ring seal is used in between the head and barrel.

Cylinder Ports
A port is an important attachment in a cylinder. It permits fluid flow both into and out of the cylinder. Further, it must be secure, as a weak port can cause a dangerous leak of hydraulic fluid under high pressure.

Usually, cylinders have standard port sizes and locations. The standard port locations are on the top of the head and cap. However, it is possible to select different locations for ports to accommodate plumbing needs or other location-specific needs. Surely, many porting options allow for a wide variety of customization to a cylinder.

On Hydraulic Cylinders, standard ports are SAE O-ring threaded ports, also referred to simply as SAE ports. SAE ports are manufactured following the SAE J1926-1 specifications. Another porting option is a four-bolt flange port. These ports are manufactured to SAE J518, code 61 and 62 specifications. Yet other port options are NPTF dry seal threads and BSPP threads.

Coating, Painting, and Polishing, Barrel
The outside surface of the barrel must be spray painted as per the relevant standards.

The internal surface of the barrel usually needs no protection with paint or chrome because the hydraulic fluid protects it from corrosion. If the hydraulic fluid is a corrosive liquid (such as water), the inside of the tube can be coated.

Coating, Piston
Hard chrome coating is provided on the piston.

Coating, Piston-rods
The surfaces of piston-rod are often treated using techniques such as Nickel-Chromium plating, laser cladding, supersonic flame spraying, or thermal spraying for making it wear-resistant and corrosion-resistant.

Coating, Further Explanation
The plating with chromium is the most common surface treatment method for the piston-rod. However, the chrome-plating can develop cracks and can cause corrosion. The corrosion resistance can be improved by having an underlying layer of nickel.

The coatings can then be finished to the desired surface roughness for optimum sealing performance.

A coating technique can be selected based on the specific operational wear conditions such as high impact or abrasive forces, corrosion, hot environment, etc.

Seals

Seals are used in a cylinder to contain the system fluid and prevent leakage. They fall into two main types. That is dynamic and static seals.

Static seals are used when there is no relative motion between mating surfaces.

Dynamic seals are used between parts that are in relative motion. Dynamic piston seals are used between the piston and cylinder bore for handling the reciprocating motion between the piston and the bore. Dynamic piston-rod seals are used between the piston-rod and head for overcoming the dynamic stresses involved.

The selection of the right seal profile and material for a given application requires consideration of many factors such as the piston-rod and bore diameters, seal groove dimensions and gaps, etc.

Assembly

The fitting section assembles all the components (rod, pistons, tubes and seals) with the necessary tools and utmost care.

Quality Control

A strict quality control procedure should be in place in every step of the production process to meet the specifications.

The quality of the hydraulic cylinder is assured by the superior manufacturing processes, the quality of raw materials, including metals and seals, used for the cylinder, the type of checks that are performed, and the expert workmanship.

Tests and Inspections, Barrel

The manufactured cylinder must undergo performance testing under load to confirm its specifications as per the requirement specifications. Ensure that all dimensions and technical requirements of the cylinder are as per the drawing. Arrange inspection by third-party agencies if required. The following tests or inspections may be carried out on the cylinder:

-Visual inspection
-Outer diameter inspection
-Bore diameter inspection
-Stroke Length
-Surface roughness inspection
-Finish test

-Pressure test
-Load test
-Hardness test
-Paint layer test
-Plating thickness (>0.04 mm)
-Concentricity inspection

-Ultrasonic test
-Penetrant test of the welds
-Destructive test for finding weld strength
-Oil port penetration inspection
-Endurance test
-Internal leakage test

Brief Explanation:

Visual inspection: Check for the presence of dust, scratches, dents, uneven painting, colour difference, sagging, and smooth plating.

Hardness: The hardness of the material can be tested in a hardness test.

Endurance test: Under the rated pressure, the tested hydraulic cylinder is continuously operated at the highest speed required by the design and continuously operated for more than 8 hours at a time.

Internal leakage test: Direct hydraulic fluid at the nominal pressure to the working chamber of the hydraulic cylinder, and measure the leakage from the working chamber to the unpressurized chamber.

External leakage test: Measure the leakage at the seal of the piston-rod and the joint.

Pressure test: In the pressure test, a cylinder is tested with a pressure of 1.5 to 2.5 times the working pressure.

15 | Objective Type Questions

1. What determines the speed of a cylinder used in a hydraulic system?
   a) Fluid flow rate
   b) System pressure
   c) Size of the applied load
   d) Stroke length

2. Wear bands in hydraulic cylinders eliminate:
   a) impact forces
   b) cylinder scoring
   c) leakage
   d) pressure drop

3. Which type of hydraulic cylinders is used to multiply force in a limited lateral space?
   a) Duplex cylinder
   b) Telescopic cylinder
   c) Tandem cylinder
   d) Ram cylinder

4. Which type of hydraulic cylinders is most suitable for the harsh service conditions?
   a) Tie-rod cylinders
   b) Mill cylinders
   c) Threaded-end cylinders
   d) Welded cylinders

16 | Review Questions

1. Define the term hydraulic actuator
2. Explain the function of a hydraulic linear actuator by taking a double-acting cylinder as an example.
3. Name two basic types of hydraulic actuators and differentiate them.
4. What are the different types of loads associated with hydraulic systems? Explain briefly.
5. Describe the relationship between the fluid flow rate, piston area and piston velocity of a hydraulic cylinder.
6. Why does a hydraulic double-acting cylinder retract at a higher speed than its speed while extending for the same input flow rate?
7. Describe the construction and design features of standard hydraulic cylinders, briefly.
8. Explain the operational and other features of different types of hydraulic cylinders.
9. Name the most general types of hydraulic cylinders.
10. What are the problems encountered by the tie-rod cylinders because of their body style?
11. What are the key advantages of welded style hydraulic cylinders?
12. Describe the constructional features of hydraulic cylinders
13. Explain the purpose of the piston-rod boot in a hydraulic cylinder.
14. Explain the purpose of the stop tube in a long-stroke hydraulic cylinder.
15. What are the different types of seal materials used in hydraulic actuators?
16. Name four different types of mounting arrangements for hydraulic cylinders?
17. Give a brief account of the commonly used hydraulic actuators.
18. Describe hydraulic actuators with neat sketches?
19. What are different methods to retract single-acting cylinders?
20. What is the difference between the single-acting and the double-acting hydraulic cylinders?

21. What is a double-rod cylinder? When is it used?
22. Why does the piston-rod of a double-acting cylinder retract at a higher velocity than when it extends for the same input flow rate?
23. What is the differential cylinder effect of a conventional hydraulic cylinder?
24. What is cylinder cushioning? Explain with a diagram.
25. Why are end cushions used in hydraulic cylinders?
26. Explain the working of a telescopic hydraulic cylinder, with a neat diagram.
27. What is a telescopic hydraulic cylinder? When is it employed?
28. Briefly explain the working of the hydraulic actuators: (1) Tandem cylinder and (2) Duplex cylinder
29. List some advantages of hydraulic cylinders.

17 | Numerical Problems

1. A 130 mm bore hydraulic cylinder is to lift a load of 720000 N. What operating pressure is required to lift the load? [Ans: 541 bar]
2. A 45 mm bore hydraulic cylinder in a construction site is used to lift a load. The gauge shows an operating pressure of 340 bar. What is the weight of the load? [Ans: 54060 N]
3. What is the speed of a hydraulic cylinder with a piston area of 132 cm^2 supplied with fluid at the rate of 100 lpm? [Ans: 0.127 m/s]
4. A single-acting hydraulic cylinder has an active area of 20.3 cm^2 and a stroke of 100 mm. How much fluid does the cylinder require per cycle? [Ans: 203 cm^3]
5. At what speed will a hydraulic cylinder with an effective area of 30 cm^2 move when powered by a pump with no load flow rate of 6000 cm^3/min? [Ans: 3.33 cm/s]
6. Calculate the forward speed of a double-acting hydraulic cylinder with a bore diameter of 100 cm and a flow rate of 100 lpm. [Ans: 2.12 mm/s]
7. A double-acting hydraulic cylinder with a single piston-rod, used for lifting and lowering heavy loads in a marine application, must produce a thrust of 80 kN and move out

with a velocity of 3 mm/s on the out-stroke. The operating pressure is 100 bar. Calculate the bore diameter of the cylinder and the flow rate required. [Ans: 10 cm, 0.024×10^{-3} m^3/s]

8. A single-acting hydraulic cylinder has a piston of 75 mm diameter and is supplied with fluid at 80 bar and a flow rate of 0.265 dm^3/s. Calculate the thrust, velocity and power. [Ans: 35360 N, 0.06 m/s, 2121.6 W]

9. A double-acting hydraulic cylinder is used to reciprocate in an application. The relief valve setting is 70 bar. The piston area is 0.0161 m^2, and the rod area is 0.00451 m^2. If the pump flow is 0.0014 m^3/s, find the speed and load-carrying capacity of the cylinder for its: (i) extension stroke and (ii) retraction stroke. [Ans: 0.087 m/s, 0.31 m/s, 112700 N, 31570 N]

10. A hydraulic cylinder has a bore dia of 8 cm and a piston-rod diameter of 4 cm. If the cylinder receives a flow of 100 lpm at 12 MPa pressure, find: a) extension as well as retraction speeds, and b) extension as well as retraction load-carrying capacities of the cylinder. [Ans: 0.33 m/s, 0.66 m/s, 60318 N, 30158 N]

11. A 7.62 cm diameter hydraulic cylinder has a 3.81 diameter piston-rod. What is the flow rate leaving through the piston-rod port of the extending cylinder, when the flow rate that enters its piston port is 30.28 lpm? [Ans: 121 lpm]

12. An 8.25 cm diameter hydraulic double-acting cylinder has a 4.445 cm diameter rod. If the cylinder receives flow at 0.0019 m^3/s and 120 bar, find the (1) thrust and pull forces produced by the cylinder, and (2) extension and retraction speeds of the cylinder, and compare. [Ans: 64800 N, 0.35 m/s, 0.5 m/s]

Objective type questions - answer key:
1-a, 2-b, 3-c, 4-b

Appendix 1

Bore diameters and Piston-rod diameters in the SI units

Table A1.1

Bore (mm)	Rod Diameter (mm)
25	12
	18
32	14
	22
40	18
	22
	28
50	22
	28
	36
63	28
	36
	45
70	40
80	36
	45
	56
90	50
100	45
	56
	70
125	56
	70
	90
150	80
160	70
	90
	110
180	100
200	90
	110
	140

Appendix 2

Theoretical Cylinder Forces in the SI Units
Force (N) = Pressure (Pascal) x Piston Area (m²)

Table A2.1

Bore (mm)	Rod dia (mm)	Force (N)	System pressure (bar)			
			69	103	138	207
32	12	Thrust	5549	8284	11099	16647
		Pull	4769	7119	9538	14307
40	16	Thrust	8671	12943	17342	26012
		Pull	7283	10872	14567	21850
50	20	Thrust	13548	20224	27096	40644
		Pull	11380	16988	22761	34141
63	20	Thrust	21509	32108	43018	64527
		Pull	19341	28872	38683	58024
80	25	Thrust	34683	51773	69366	104050
		Pull	31296	46717	62592	93888
100	32	Thrust	54192	80896	108385	162577
		Pull	48643	72612	97286	145929
125	32	Thrust	84676	126400	169351	254027
		Pull	79126	118116	158252	237379
160	40	Thrust	138733	207094	277465	416198
		Pull	130062	194150	260124	390186
200	40	Thrust	216770	323584	433540	650310
		Pull	208099	310641	416198	624297

Example

Bore diameter = 32 mm = 0.032 m

Rod diameter = 12 mm = 0.012 m

Pressure, P = 69 bar = 69 x 10^5 Pa

Piston area, A_{ext} = $\prod D^2/4$ = 3.1416 x $0.032^2/4$ = 0.0008042 m²

Effective area for pull stroke, A_{ret} = $\prod (D^2 - d^2)/4$

= 3.14 $(0.032^2 - 0.012^2)/4$ = 0.0006908 m²

Thrust = P x Aext = 69 x 10^5 x 0.0008042 = 5549 N

Pull = P x Aret = 69 x 10^5 x 0.0006908 = 4767 N

Appendix 3

Different types of Piston Seals
- Lip Seal
- Spring-Loaded PTFE Seal
- Magnetic Piston, stainless steel cylinder body, single bi-directional piston seal
- Magnetic Piston, carbon steel body, single bi-directional piston seal
- Magnetic Piston, Aluminum Tube

Ports
- SAE Straight Thread O-Ring
- NPTF Ports (Dry Seal Pipe Thread)
- BSP Ports (Parallel Thread ISO 228)
- BSPT Ports (Taper Thread)
- Metric Thread Ports
- Metric Thread Ports per ISO 6149

Mounting Styles
- Tie Rods Extended Head-end
- Tie Rods Extended Cap-end
- Tie Rods Extended at Both Ends
- Head Rectangular Flange
- Head Square Flange
- Cap Rectangular Flange
- Cap Square Flange
- Side Lug
- Side Tapped
- Cap Fixed Clevis
- Head Trunnion
- Cap Trunnion
- Intermediate Fixed Trunnion
- Spherical Bearing

Appendix 4

Important specifications to be considered while selecting hydraulic cylinders

Table A4.1

Bore size	Should satisfy the requirement of thrust/pull for the operating pressure
Piston-rod diameter	Should prevent piston rod buckling
Single rod / Double rod	
Cushions	Yes / No If yes, Head-end, Cap-end, or both ends?
Stop tube	Yes / No
Piston and piston-rod Seal type	Fluid and temperature compatibility?
Stroke length	
Piston-rod-end thread style	
Port size	For a given speed requirement
Port position	
Mounting style	
Piston rod and mounting accessories	To attach the cylinder to the load
Optional accessories	
Fluid medium	

Appendix 5

Seal Materials and their Temperature Ranges

Table A5.1

Material	Temperature Range
Nitrile	-30°C to 100°C
H-Nitrile (Hydrogenated Nitrile)	-35°C to 150°C
Viton	-25°C to 262°C
Silicone	-60°C to 232°C
EPDM	-45°C to 150°C
Polyurethane	-30°C to 110°C
Nylon	-40°C to 120°C
Teflon, virgin	-200°C to 260°C
Teflon, filled	-200°C to 260°C

18 | References

1. Andrew Parr, Hydraulics & Pneumatics, A technician's and engineer's guide, 2nd Edition, Butterworth, Heinemann, 1998
2. Anthony Esposito, Fluid Power with Applications, 6th Edition, Prentice-Hall of India, 2006
3. Article on '8 Steps You Should Follow To Manufacturer A High-quality Hydraulic Cylinder', by Antony, Taizhou Chuanhu Hydraulic Machinery Co., Ltd, Zhejiang Province, China
4. Article on 'A Brief Look At The Hydraulic Cylinder Manufacturing Process', Ranger Caradoc Hydraulics Ltd., West Bromwich, United Kingdom
5. Article on 'Clamping elements', HYDROKOMP®, Mücke, Germany, www.hydrokomp.de
6. Article on 'Custom Hydraulic Cylinder Manufacturing', CATSU Hydraulic Machinery Equipment Co., Ltd.
7. Article on 'Cylinders-Part 1', Hydraulics & Pneumatics Magazine, Penton Media, Inc
8. Article on 'FLUID POWER Design Data Sheet' published by WOMAC EDUCATIONAL PUBLICATIONS, Womac Machine Supply Co., 13835 Senlac Dr Farmers Branch, TX 75234
9. Article on 'How to Determine what kind of Porting you will need for your Hydraulic Cylinder Application'' SHEFFER PNEUMATIC AND HYDRAULIC CYLINDERS
10. Article on 'HYDRAULIC CLAMPING', VEKTEK LLC, U.S.A., www.vektek.com
11. Article on 'Hydraulic cylinder anatomy', by Frank J. Bartos, Control Engineering Resource Center
12. Article on 'Hydraulic cylinder', HYDRAULIC EQUIPMENT & TOOLS MARKETPLACE, Hydraulic Equipment Manufacturers
13. Article on 'Hydraulic Cylinders', INTEGRAL HYDRAULIK GmbH & Co. KG Hanns-Martin-Schleyer-Str. 20, 47877 Willich Postfach 500 209

14. Article on 'Hydraulic swing clamp', KOSMEK, Japan
15. Article on 'Hydraulic Swing Cylinders – Top Mounted, Double Acting', Precision Engineering Accessories, Bangalore, India, www.preacindia.com
16. Article on 'Industrial hydraulic cylinders', Herbert Hänchen GmbH
17. Article on 'Mini Swing Clamps with Sturdy Swing Mechanism, threaded-body type, double acting, max. operating pressure 150 bar', Issue B1.848/12-10E, ROEMHELD GmbH
18. Article on 'Requirements and material section of hydraulic cylinder barrel', Wuxi Longzhichen Machinery Co., Ltd., https://www.honedtubing.com/info/requirements-and-material-section-of-hydraulic-18363999.html
19. Articles on 'Hydraulic Cylinder Seal Selection', 'Hydraulic Cylinder Repair Tutorial', 'The Advantages of Hydraulic Cylinders', 'Hydraulic Cylinders- Design Considerations for Hydraulic Cylinders', 'Hydraulic Cylinders-Engineering and Design Tips', 'Hydraulic Oil Tutorial, Hydraulic Cylinder Safety Tutorial', and 'Advantages of Welded Body Hydraulic Cylinders', HYCO ULTRAMETAL, Kitchener, Strasburg Road, Unit C Kitchener, Ontario
20. Brochure on 'Cylinders Capabilities', Document No. M-CYOV-MR001-E, January 2007, Eaton, USA
21. Catalogue on 'High-pressure hydraulics', EURO PRESS PACK, Via M. Disma, Carasco (GE), ITALY
22. Catalogue on 'Hydraulic cylinder, Type CDL1, Series 1X, Nominal pressure 160 bar (16 MPa), RE 17325/09.05', Bosch Rexroth AG Hydraulics
23. Catalogue on 'Medium Duty Hydraulic Cylinders, Series 3L', 4/10 / No. HY08-1130-1/NA, Parker Hannifin Corporation, Cylinder Division, USA. www.parker.com/cylinder
24. Catalogue on 'Metric Hydraulic Cylinders, Series HMI, Conforms to ISO 6020/2 (1991), for working pressures up to 210 bar', Parker Hannifin Corporation, Cylinder Division, USA

25. Catalogue on Series H, Milwaukee Cylinder, Milwaukee, USA, www.milwaukeecylinder.com
26. Document on 'OVERVIEW AND SELECTION GUIDE' Doc. no. 941186/Mat. No. 267436 EN, Balluff Inc. 8125 Holton Drive Florence, KY 41042
27. Hydraulic Cylinder Troubleshooting - Parker Hannifin. (n.d.). Retrieved from http://parker.com/literature/Literature%20Files/euro_cylinder/v4/Trouble_1242-1-gb.pdf_br
28. Vocational Training Course, HYDRAULICS – 21 Exercises with Instructions, published by Bundesinstitut fur Berufsbildungsforschung, Berlin, 1973
29. William D. Wolansky et al., Fundamentals of Fluid power, Houghton Mifflin Company, Boston, 1977

Fluid Power Educational Series Books

1. Pneumatic Systems and Circuits -Basic Level (In the SI Units)
2. Industrial Pneumatics -Basic Level (In the English Units)
3. Pneumatic Systems and Circuits -Advanced Level
4. Electro-Pneumatics and Automation
5. Design of Pneumatic Systems (In the SI Units)
6. Design Concepts in Pneumatic Systems (In the English Units)
7. Maintenance, Troubleshooting, and Safety in Pneumatic Systems
8. Industrial Hydraulic Systems and Circuits -Basic Level (In the SI Units)
9. Industrial Hydraulics -Basic Level (In the English Units)
10. Hydraulic Fluids
11. Hydraulic Filters: Construction, Installation Locations, and Specifications
12. Hydraulic Power Packs (In the SI Units)
13. Power Packs in Hydraulic Systems (In the English Units)
14. Hydraulic Cylinders (In the SI Units)
15. Hydraulic Linear Actuators (In the English Units)
16. Hydraulic Motors (In the SI Units)
17. Hydraulic Rotary Actuators (In the English Units)
18. Hydraulic Accumulators and Circuits (In the SI Units)
19. Accumulators in Hydraulic Systems (In the English Units)
20. Hydraulic Pipes, Tubes, and Hoses (In the SI Units)
21. Pipes, Tubes, and Hoses in Hydraulic Systems (In the English Units)
22. Design of Industrial Hydraulic Systems (In the SI Units)
23. Design Concepts in Industrial Hydraulic Systems (In the English Units)
24. Maintenance, Troubleshooting, and Safety in Hydraulic Systems
25. Hydrostatic Transmissions (HSTs) (In the SI Units)
26. Concepts of Hydrostatic Transmissions (In the English Units)

27. Load Sensing Hydraulic Systems (In the SI Units)
28. Concepts of Load Sensing Hydraulic Systems (In the English Units)
29. Electro-hydraulic Proportional Valves
30. Electro-hydraulic Servo Valves
31. Cartridge Valves
32. Electro-hydraulic Systems and Relay Circuits
33. Practical Book: Pneumatics - Basic Level
34. Practical Book: Electro-pneumatics - Basic Level
35. Practical Book: Industrial Hydraulics – Basic Level
36. Programmable Logic Controllers and Programming Concepts
37. Compressed Air Dryers

For more details, please visit: **https://jojibooks.com**

About the Author

Joji Parambath is a trainer in the field of Pneumatics, Hydraulics, and PLC, for over 25 years. During his career, he has trained numerous professionals from the industries as well as faculty members and students of engineering institutions.

At present, he is the key trainer at Fluidsys Training Centre, Bangalore, India, (https://fluidsys.org) which is providing training in the field of Pneumatics and Hydraulics. He has already written two books on Pneumatics and Hydraulics. The publication of the present series of 32 books is intended to restructure and update the existing books.

The author wishes to thank all trainees for their lively interaction and many useful suggestions during the training programmes that prompted the author to write the present series of books. You may send your feedback to joji.p@hotmail.com

10th June 2020